U0187452

普通高等教育系列教材

计算机专业英语

第3版

张强华　司爱侠　朱丽丽　张千帆　编著

机 械 工 业 出 版 社

本书涵盖与计算机专业相关的基础知识、基本概念及发展动向等方面的英文知识。全书有12个单元，每个单元都包括课文、单词、词组、缩略语、难句讲解、构词法、习题、参考译文（Text A）。附录中的单词表、词组表和缩略语表既可供读者复习和记忆单词，也可作为小词典长期使用。

本书还提供了丰富的栏目："Career Training"栏目，直通职场，有助于读者提升目前IT行业注重的能力；课文听力材料，扫二维码即可收听；"软件水平考试试题解析"通过对历年典型试题的解析、讲解，兼顾行业考试，以例题点拨思路，以习题训练热身，使读者掌握应试要领。

本书既可作为高等院校信息类专业（包括计算机科学与技术、网络工程、软件工程、信息管理、电子商务、电子政务等）的英语教材，也可供参加计算机行业各种考试的读者备考之用，还可作为培训班教材和从业人员自学用书。

图书在版编目（CIP）数据

计算机专业英语 / 张强华等编著. —3版. —北京：机械工业出版社，2020.8（2023.6重印）
普通高等教育系列教材
ISBN 978-7-111-66019-4

Ⅰ. ①计… Ⅱ. ①张… Ⅲ. ①电子计算机－英语－高等学校－教材
Ⅳ. ①TP3

中国版本图书馆CIP数据核字（2020）第118724号

机械工业出版社（北京市百万庄大街22号 邮政编码100037）
策划编辑：胡 静 责任编辑：胡 静
责任校对：张艳霞 责任印制：单爱军
北京虎彩文化传播有限公司印刷

2023年6月第3版·第10次印刷
184mm×260mm·17.75印张·438千字
标准书号：ISBN 978-7-111-66019-4
定价：59.90元

电话服务
客服电话：010-88361066
010-88379833
010-68326294

网络服务
机 工 官 网：www.cmpbook.com
机 工 官 博：weibo.com/cmp1952
金 书 网：www.golden-book.com
机工教育服务网：www.cmpedu.com

前　言

科技兴则民族兴，科技强则国家强。党的二十大报告指出，必须坚持科技是第一生产力、人才是第一资源、创新是第一动力，开辟发展新领域新赛道，不断塑造发展新动能新优势。

IT 领域的一个主要特征就是发展极快，新技术层出不穷、软件频繁升级、设备不断更新。要跟上其发展步伐，需要不断地学习。英语能力关乎从业人员的核心竞争力，因此专业人士都在竭尽全力地提高自己的专业英语水平，打造强劲工作能力。各所高校也都把专业英语作为必修课，以提高学生跨越行业门槛的竞争能力。

本书自第 1 版出版以来，深受各类读者的喜爱，得到许多教师的认可与推荐，被多所院校选为教材。本书作为面向职场、贴近行业、强化能力的一本改版教材，其特色如下：

1．贴近行业选择素材，面向职场强化能力

依据当前 IT 行业的最新发展，精心选择素材。所选课文既包括行业的基础知识和基本概念，也包括常用软件及常用设备，同时注意未来发展和动向。本次修订增加了大数据、数据挖掘、云计算、人工智能和虚拟现实等新知识。

同时，根据 IT 行业当前特别注重能力的特点，特别增加了“Career Training”栏目，内容包括“如何写求职信”“如何写简历”及“面试技巧”等内容，使得本书具有“直通职场”的特征。

2．有效进行听力训练，告别“无声”教材时代

在当前 IT 行业的工作场合，人们越来越多地需要沟通。例如，听外国专家的技术报告、与国外合作伙伴进行技术交流等都要具有听说能力，甚至面试时就需要听说交流。现在大多数教材无视职场的这一要求，未提供听力训练，而本书精心设计了听力内容，配备了听力材料，读者通过扫描二维码即可收听，这将有力地提升读者运用英语的能力。

3．突出教材特征，易学易教易用

本书体例上以“Unit”为单位，每个“Unit”由以下几部分组成：课文——选材得当、软硬件并重；单词——给出课文中出现的新词，读者由此可以积累计算机专业的基本词汇；词组——给出课文中的常用词组；缩略语——给出课文中出现的、业内人士必须掌握的缩略语；难句讲解——讲解课文中出现的疑难句子，培养读者的阅读理解能力；构词法——既可以帮助读者记忆单词，也可以帮助读者“破解”新出现的词汇；习题——既有针对课文的练习，也有针对行业考试的练习；参考译文（Text A）——供读者对照学习。附录中的单词表、词组表和缩略语表既可供读者复习和记忆单词之用，也可作为小词典长期使用。

4．针对相关考试，解析讲解训练

本书既考虑教学需要，也兼顾 IT 行业的一些考试（如软件水平考试），精选了历年的典型试题，通过解析、讲解，使读者掌握应试要领。同时，提供了代表性真题以供读者练习。以例题点拨思路，以习题训练热身。

5．浓缩多年经验，提供后续支持

本书作者已经出版了 8 部计算机英语教材，有 10 余年相关经验；熟悉学生情况、熟知学生毕业后的就业环境。

本书配有教学课件，需要的教师可登录 www.cmpedu.com 免费注册，审核通过后下载，或联系编辑索取（微信：15910938545，电话：010-88379739）。

本书由张强华、司爱侠、朱丽丽和张千帆编著。

由于编者水平有限，书中难免存在疏漏之处，敬请读者批评指正。

编　者

目　录

Unit 1

Text A

Computer Hardware

A computer is a machine that can be instructed to carry out sequences of arithmetic or logical operations automatically via computer programming. Modern computers have the ability to follow generalized sets of operations, called programs. These programs enable computers to perform an extremely wide range of tasks. A "complete" computer including the hardware, the operating system, and peripheral equipment required and used for "full" operation can be referred to as a computer system.[1] This term may as well be used for a group of computers that are connected and work together, in particular a computer network or computer cluster.

Computers are used as control systems for a wide variety of industrial and consumer devices. This includes simple special purpose devices like microwave ovens and remote controls, factory devices such as industrial robots and computer aided design, and also general purpose devices like personal computers and mobile devices such as smartphones. The Internet is run on computers and it connects hundreds of millions of other computers and their users.

Conventionally, a modern computer consists of at least one processing element, typically a central processing unit (CPU), and some form of memory. The processing element carries out arithmetic and logical operations, and a sequencing and control unit can change the order of operations in response to stored information. Peripheral devices include input devices (keyboards, mouse, etc.), output devices (monitor screens, printers, etc.). Peripheral devices allow information to be retrieved from an external source and they enable the result of operations to be saved and retrieved.

The term hardware covers all of those parts of a computer that are tangible physical objects. Circuits, computer chips, graphic cards, sound cards, memory (RAM), motherboards, displays, power supplies, cables, keyboards, printers and mouse are all hardware.

A general purpose computer has four main components: the arithmetic and logic unit (ALU), the control unit, the memory, and the input and output devices (collectively termed I/O). These parts are interconnected by buses, often made of groups of wires. Inside each of these parts are thousands to trillions of small electrical circuits which can be turned off or on by means of an electronic switch. Each circuit represents a bit (binary digit) of information so that when the circuit is on it represents a

"1", and when off it represents a "0" . The circuits are arranged in logic gates so that one or more of the circuits may control the state of one or more of the other circuits.

1. Input devices

When unprocessed data is sent to the computer with the help of input devices, the data is processed and sent to output devices. The input devices may be hand-operated or automated. The act of processing is mainly regulated by the CPU. Some examples of input devices are computer keyboard, digital camera, digital video, graphics tablet, image scanner, microphone, mouse, touch screen.

2. Output devices

The means through which computer gives output are known as output devices. Some examples of output devices are computer monitor, printer, PC speaker, sound card, video card.

3. Control unit

The control unit (often called a control system or central controller) manages the computer's various components; it reads and interprets (decodes) the program instructions, transforming them into control signals that activate other parts of the computer. Control systems in advanced computers may change the order of execution of some instructions to improve performance.

A key component common to all CPUs is the program counter, a special memory cell (a register) that keeps track of which location in memory the next instruction is to be read from.[2]

The control system's functions are as follows. Note that this is a simplified description, and some of these steps may be performed concurrently or in a different order depending on the type of CPU:

(1) Read the code for the next instruction from the cell indicated by the program counter.

(2) Decode the numerical code for the instruction into a set of commands or signals for each of the other systems.

(3) Increase the program counter so it points to the next instruction.

(4) Read whatever data the instruction requires from cells in memory (or perhaps from an input device). The location of this required data is typically stored within the instruction code.

(5) Provide the necessary data to an ALU or register.

(6) If the instruction requires an ALU or specialized hardware to complete, instruct the hardware to perform the requested operation.

(7) Writes the result from the ALU back to a memory location or to a register or perhaps an output device.

(8) Jump back to step (1).

4. Central processing unit (CPU)

The control unit, ALU, and registers are collectively known as a central processing unit (CPU). Early CPUs were composed of many separate components but since the mid-1970s CPUs have typically been constructed on a single integrated circuit called a microprocessor.

5. Arithmetic and logic unit (ALU)

The ALU is capable of performing two classes of operations: arithmetic and logic. The set of arithmetic operations that a particular ALU supports may be limited to addition and subtraction, or might include multiplication, division, trigonometric functions such as sine, cosine, etc., and square roots. Some can only operate on integers while others use floating point to represent real numbers, albeit with limited precision. However, any computer that is capable of performing just the simplest operations can be programmed to break down the more complex operations into simple steps that it can perform. Therefore, any computer can be programmed to perform any arithmetic operation, although it will take more time to do so if its ALU does not directly support the operation.[3] An ALU may also compare numbers and return boolean values (true or false) depending on whether one is equal to, greater than or less than the other. Logic operations involve Boolean logic: AND, OR, XOR, and NOT. These can be useful for creating complicated conditional statements and processing boolean logic.

Superscalar computers may contain multiple ALUs, allowing them to process several instructions simultaneously. Graphics processors and computers with SIMD and MIMD features often contain ALUs that can perform arithmetic on vectors and matrices.

6. Memory

A computer's memory can be viewed as a list of cells into which numbers can be placed or read. Each cell has a numbered "address" and can store a single number. The information stored in memory may represent practically anything. Letters, numbers, even computer instructions can be placed into memory with equal ease. Since the CPU does not differentiate between different types of information, it is the software's responsibility to give significance to what the memory sees as nothing but a series of numbers.

In almost all modern computers, each memory cell is set up to store binary numbers in groups of eight bits (called a byte). Each byte is able to represent 256 different numbers ($2^8 = 256$); either from 0 to 255 or −128 to +127. To store larger numbers, several consecutive bytes may be used (typically, two, four or eight). When negative numbers are required, they are usually stored in two's complement notation. A computer can store any kind of information in memory if it can be represented numerically.

The CPU contains a special set of memory cells called registers that can be read and written to much more rapidly than the main memory area. There are typically between two and one hundred registers depending on the type of CPU. Registers are used for the most frequently needed data items to avoid having to access main memory every time data is needed. As data is constantly being worked on, reducing the need to access main memory (which is often slow compared to the ALU and control units) greatly increases the computer's speed.

Computer main memory comes in two principal varieties:

- random-access memory (RAM).
- read-only memory (ROM).

RAM can be read and written to anytime the CPU commands it, but ROM is preloaded with data and software that never changes, therefore the CPU can only read from it. ROM is typically used to store the computer's initial start-up instructions. In general, the contents of RAM are erased when the power to the computer is turned off, but ROM retains its data indefinitely. In a PC, the ROM contains a specialized program called the BIOS that orchestrates loading the computer's operating system from the hard disk drive into RAM whenever the computer is turned on or reset.[4]

In embedded computers, which frequently do not have disk drives, all of the required software may be stored in ROM. Software stored in ROM is often called firmware, because it is notionally more like hardware than software. Flash memory blurs the distinction between ROM and RAM, as it retains its data when the computer is turned off but is also rewritable.

7. Multiprocessing

Some computers are designed to distribute their work across several CPUs in a multiprocessing configuration, a technique once employed only in large and powerful machines such as supercomputers, mainframe computers and servers. Multiprocessor and multi-core (multiple CPUs on a single integrated circuit) personal and laptop computers are now widely available.

Supercomputers in particular often have highly unique architectures that differ significantly from the basic stored-program architecture and from general purpose computers. They often feature thousands of CPUs, customized high-speed interconnects, and specialized computing hardware. Such designs tend to be useful only for specialized tasks. Supercomputers usually see usage in large-scale simulation, graphics rendering, and cryptography applications.

New Words

computer	[kəm'pju:tə]	*n.*（电子）计算机，电脑
hardware	['hɑ:dweə]	*n.*硬件
logical	['lɒdʒɪkl]	*adj.*逻辑的；符合逻辑的
operation	[ˌɒpə'reɪʃn]	*n.*运算；操作，经营
program	['prəʊgræm]	*n.*程序
		*v.*编写程序
set	[set]	*n.*集合
		*vt.*设置
perform	[pə'fɔ:m]	*v.*执行
task	[tɑ:sk]	*n.*工作，任务
complete	[kəm'pli:t]	*adj.*完整的；完全的；完成的
		*vt.*完成，使完满
peripheral	[pə'rɪfərəl]	*adj.*外围的；次要的
		*n.*外部设备
equipment	[ɪ'kwɪpmənt]	*n.*设备，装备
network	['netwɜ:k]	*n.*网络
industrial	[ɪn'dʌstrɪəl]	*adj.*工业的，产业的
device	[dɪ'vaɪs]	*n.*装置，设备
remote	[rɪ'məʊt]	*adj.*远程的
		*n.*远程操作
robot	['rəʊbɒt]	*n.*机器人；遥控装置
mobile	['məʊbaɪl]	*adj.*可移动的
smartphone	[smɑ:tfəʊn]	*n.*智能手机

Internet	[ɪn'tənet]	*n.*因特网
process	['prəʊses]	*vt.*加工；处理
		*n.*过程
memory	['memərɪ]	*n.*存储器，内存
store	[stɔ:]	*v.*（在计算机里）存储；贮存
information	[ˌɪnfə'meɪʃn]	*n.*信息；数据；消息
keyboard	['ki:bɔ:d]	*n.*键盘
mouse	[maʊs]	*n.*鼠标器
screen	[skri:n]	*n.*屏幕
printer	['prɪntə]	*n.*打印机
tangible	['tændʒəbl]	*adj.*可触知的；确实的，真实的；实际的
chip	[tʃɪp]	*n.*芯片
motherboard	['mʌðəbɔ:d]	*n.*主板，母板
display	[dɪ'spleɪ]	*vi.*（计算机屏幕上）显示
		*n.*显示，显示器
bus	[bʌs]	*n.*总线
wire	['waɪə]	*n.*电线
bit	[bɪt]	*n.*位，比特（二进制数字中的位，信息量的度量单位）
unprocessed	[ʌn'prəʊsest]	*adj.*未处理的
automate	['ɔ:təmeɪt]	*vt.*自动化，使自动操作
regulate	['regjuleɪt]	*vt.*控制，管理；调节，调整；校准
scanner	['skænə]	*n.*扫描器；扫描设备
touch screen	['tʌtʃskri:n]	*n.*触摸屏
controller	[kən'trəʊlə]	*n.*控制器
interpret	[ɪn'tɜ:prɪt]	*vt.*解释
		*vi.*做解释
register	['redʒɪstə]	*n.*寄存器
description	[dɪ'skrɪpʃn]	*n.*描述，形容；种类，类型
concurrently	[kən'kʌrəntlɪ]	*adv.*同时地
code	[kəʊd]	*n.*代码
		*vt.*编码
indicate	['ɪndɪkeɪt]	*vt.*表明，标示，指示
instruct	[ɪn'strʌkt]	*vt.*指示，通知；命令
conceptually	[kən'septʃuəlɪ]	*adv.*概念地
microprocessor	[ˌmaɪkrəʊ'prəʊsesə]	*n.*微处理器
integer	['ɪntɪdʒə]	*n.*整数
boolean	['bu:lɪən]	*adj.*布尔的
conditional	[kən'dɪʃənl]	*adj.*条件的，假定的

superscalar	['su:pəˌskeɪlə]	*n*.超标量结构
vector	['vektə]	*n*.向量，矢量
matrix	['meɪtrɪks]	*n*.矩阵
address	[ə'dres]	*n*.地址
responsibility	[rɪˌspɒnsə'bɪlətɪ]	*n*.责任，职责
significance	[sɪg'nɪfikəns]	*n*.意义，意思；重要性
byte	[baɪt]	*n*.字节
notation	[nəʊ'teɪʃn]	*n*.标记符号，表示法
preload	[ˌpri: 'ləʊd]	*n*.& *v*.预载，预先装载
start-up	['stɑ:t-ʌp]	*n*.启动
erase	[ɪ'reɪz]	*vt*.擦掉；抹去；清除
orchestrate	['ɔ:kɪstreɪt]	*vt*.精心策划
embedded	[ɪm'bedɪd]	*adj*.嵌入的，植入的，内含的
rewritable	[ˌri: 'raɪtəbl]	*adj*.可重写的，可复写的
multiprocessing	['mʌltɪˌprəʊsesɪŋ]	*n*.多重处理，多处理（技术）
supercomputer	['su:pəkəmpju:tə]	*n*.超级计算机，巨型计算机
multi-core	['mʌlti-kɔ:]	*adj*.多核的
architecture	['ɑ:kɪtektʃə]	*n*.体系结构
significantly	[sɪg'nɪfikəntlɪ]	*adv*.意味深长地；值得注目地
interconnect	[ˌɪntəkə'nekt]	*vi*.互相连接，互相联系 *vt*.使互相连接；使互相联系
cryptography	[krɪp'tɒgrəfɪ]	*n*.密码学，密码术

Phrases

be instructed to	被指令做某事
carry out	执行，进行，完成
operating system	操作系统
peripheral equipment	外围设备，外部设备
be referred to as	被称作，被称为
a group of	一群，一组
a wide variety of	种种，多种多样
microwave oven	微波炉
computer-aided design	计算机辅助设计
personal computer	个人计算机
mobile device	移动设备，移动装置
consists of...	由……组成
in response to...	对……做出反应
input device	输入设备
output device	输出设备

be retrieved from...	从……收集，从……获取
computer chip	计算机芯片
graphic card	图形卡，显示卡
sound card	音效卡，声卡
power supply	电源；供电
control unit	控制单元，控制器
electrical circuit	电路
turn off	关闭
turn on	打开
by means of	借助……手段，依靠……方法
electronic switch	电子开关
binary digit	二进制位，二进制数字
logic gate	逻辑门
graphics tablet	绘图板；图形输入板
memory cell	存储单元
keep track of	跟踪
integrated circuit	集成电路
trigonometric function	三角函数
square root	平方根
floating-point	浮点，浮点法
real number	实数
boolean value	布尔值
conditional statement	条件语句
be viewed as	被看作是，视为
set up	建立；准备；安排
binary number	二进制数
negative number	负数
hard disk drive	硬盘驱动器
flash memory	闪存
be restricted to	仅限于
be designed to…	被设计为……
mainframe computer	大型计算机
graphics rendering	图形渲染

Abbreviations

CPU (Central Processing Unit)	中央处理器
RAM (Random Access Memory)	随机存取存储器
ALU (Arithmetic and Logic Unit)	算术逻辑部件
I/O (Input/Output)	输入/输出

PC (Personal Computer) 个人计算机
SIMD (Single Instruction Multiple Data) 单指令多数据
MIMD (Multiple Instruction Multiple Data) 多指令多数据
ROM (Read-Only Memory) 只读存储器
BIOS (Basic Input/Output System) 基本输入/输出系统

Notes

[1] A "complete" computer including the hardware, the operating system, and peripheral equipment required and used for "full" operation can be referred to as a computer system.

本句中，A "complete" computer 作主语，can be referred to 作谓语。including the hardware, the operating system, and peripheral equipment required and used for "full" operation 是一个现在分词短语作定语，修饰和限定主语 A "complete" computer。在该现在分词短语中，required and used for "full" operation 是过去分词短语作定语，修饰和限定 the hardware, the operating system, and peripheral equipment。

[2] A key component common to all CPUs is the program counter, a special memory cell (a register) that keeps track of which location in memory the next instruction is to be read from.

本句中，a special memory cell (a register) that keeps track of which location in memory the next instruction is to be read from 是一个名词短语，对 the program counter进行补充说明。在该名词短语中，that keeps track of which location in memory the next instruction is to be read from 是一个定语从句，修饰和限定 a special memory cell。keep track of 的意思是"跟踪，追踪"。

[3] Therefore, any computer can be programmed to perform any arithmetic operation, although it will take more time to do so if its ALU does not directly support the operation.

本句中，although it will take more time to do so if its ALU does not directly support the operation 是一个让步状语从句，修饰谓语 can be programmed。在该让步状语从句中，if its ALU does not directly support the operation 是一个条件状语从句，修饰从句的谓语 will take more time to do so。

[4] In a PC, the ROM contains a specialized program called the BIOS that orchestrates loading the computer's operating system from the hard disk drive into RAM whenever the computer is turned on or reset.

本句中，called the BIOS是一个过去分词短语作定语，修饰和限定 a specialized program。that orchestrates loading the computer's operating system from the hard disk drive into RAM whenever the computer is turned on or reset 是一个定语从句，修饰和限定 the BIOS。whenever the computer is turned on or reset 是一个时间状语从句，修饰定语从句的谓语 orchestrates。

Text B

Text B
John Von Neumann

John Von Neumann

John Von Neumann (Figure 1-1) was a child prodigy, born into a banking family in Budapest,

Hungary. When only six years old he could divide eight-digit numbers in his head. He received his early education in Budapest, under the tutelage of M. Fekete, with whom he published his first paper at the age of 18. Entering the University of Budapest in 1921, he studied Chemistry, moving his base of studies to both Berlin and Zurich before receiving his diploma in 1925 in Chemical Engineering. He returned to his first love of mathematics in completing his doctoral degree in 1928. He quickly gained a reputation in set theory, algebra, and quantum mechanics. At a time of political unrest in central Europe, he was invited to visit Princeton University in 1930, and when the Institute for Advanced Studies was founded there in 1933, he was appointed to be one of the original six Professors of Mathematics, a position which he retained for the remainder of his life. At the instigation and sponsorship of Oskar Morganstern, Von Neumann and Kurt Gödel became US citizens in time for their clearance for wartime work. There is an anecdote which tells of Morganstern driving them to their immigration interview, after having learned about the US Constitution and the history of the country. On the drive there Morganstern asked them if they had any questions which he could answer. Gödel replied that he had no questions but he had found some logical inconsistencies in the Constitution that he wanted to ask the Immigration officers about. Morganstern strongly recommended that he not ask questions, just answer them!

Figure 1-1 John Von Neumann

Form 1936 to 1938 Alan Turing was a graduate student in the Department of Mathematics at Princeton and did his dissertation under Alonzo Church. Von Neumann invited Turing to stay on at the Institute as his assistant but he preferred to return to Cambridge; a year later Turing was involved in war work at Bletchley Park. This visit occurred shortly after Turing's publication of his 1934 paper "On Computable Numbers with an Application to the Entscheidungs-problem" which involved the concepts of logical design and the universal machine. It must be concluded that Von Neumann knew of Turing's ideas, though whether he applied them to the design of the IAS Machine ten years later is questionable.

Von Neumann's interest in computers differed from that of his peers by his quickly perceiving the application of computers to applied mathematics for specific problems, rather than their mere application to the development of tables. During the war, Von Neumann's expertise in hydrodynamics, ballistics, meteorology, game theory, and statistics was put to good use in several projects. This work led him to consider the use of mechanical devices for computation, and although the stories about Von Neumann imply that his first computer encounter was with the ENIAC, in fact it was with Howard Aiken's Harvard Mark I calculator. His correspondence in 1944 shows his interest with the work of not only Aiken but also the electromechanical relay computers of George Stibitz, and the work by Jan Schilt at the Watson Scientific Computing Laboratory at Columbia University. During the latter years of World War II, Von Neumann was playing the part of an executive management consultant, serving on several national committees, applying his amazing ability to rapidly see through problems to their solutions. Through this means he was also a conduit between groups of scientists who were otherwise shielded from each other by the requirements of secrecy. He brought together the needs of the Los Alamos National Laboratory (and the Manhattan Project) with the capabilities of firstly the engineers at the Moore School of Electrical Engineering who were building the ENIAC, and later his own work on building the IAS machine. Several "supercomputers" were built by National Laboratories as copies of his machine.

Postwar Von Neumann concentrated on the development of the Institute for Advanced Studies (IAS) computer and its copies around the world. His work with the Los Alamos group continued and he continued to develop the synergism between computers capabilities and the needs for computational solutions to nuclear problems related to the hydrogen bomb.

Any computer scientist who reviews the formal obituaries of John Von Neumann of the period shortly after his death will be struck by the lack of recognition of his involvement in the field of computers and computing. His Academy of Sciences Biography, written by Salomon Bochner [1958], for example, includes but a single, short paragraph in ten pages "... in 1944 Von Neumann's attention turned to computing machines and, somewhat surprisingly, he decided to build his own. As the years progressed, he appeared to thrive on the multitudinousness of his tasks. It has been stated that Von Neumann's electronic computer hastened the hydrogen bomb explosion on November 1, 1952." Dieudonné [1981] is a little more generous with words but appears to confuse the concept of the stored program concept with the wiring of computers: "Dissatisfied with the computing machines available immediately after the war, he was led to examine from its foundations the optimal method that such machines should follow, and he introduced new procedures in the logical organization, the 'codes' by which a fixed system of wiring could solve a great variety of problems"!

From the point of view of Von Neumann's contributions to the field of computing, including the application of his concepts of mathematics to computing, and the application of computing to his other interests such as mathematical physics and economics, perhaps the most comprehensive is by Herman Goldstine [1972]. There has been some criticism of Goldstine's perspective since he personally was intimately involved in Von Neumann's computing activities from the time of their chance meeting on the railroad platform at Aberdeen in 1944 through their joint activities at the Institute for Advanced Studies in developing the IAS machine.

There is no doubt that his insights into the organization of machines led to the infrastructure which is now known as the "Von Neumann Architecture". However, Von Neumann's ideas were not along those lines originally; he recognized the need for parallelism in computers but equally well recognized the problems of construction and hence settled for a sequential system of implementation. Through the report entitled First Draft of a Report on the EDVAC [1945], authored solely by Von Neumann, the basic elements of the stored program concept were introduced to the industry. A retrospective examination of the development of this idea reveals that the concept was discussed by J. Presper Eckert, John Mauchly, Arthur Burks, and others in connection with their plans for a successor machine to the ENIAC. The "Draft Report" was just that, a draft, and although written by Von Neumann, was intended to be the joint publication of the whole group. The EDVAC was intended to be the first stored program computer, but the summer school at the Moore School in 1946 there was so much emphasis in the EDVAC that Maurice Wilkes, Cambridge University Mathematical Laboratory, conceived his own design for the EDSAC, which became the world's first operational, production, stored-program computer.

In the 1950s Von Neumann was employed as a consultant to IBM to review proposed and ongoing advanced technology projects. One day a week, Von Neumann "held court" at 590 Madison Avenue, New York. On one of these occasions in 1954 he was confronted with the FORTRAN concept; John Backus remembered Von Neumann being unimpressed and that he asked "why would you want more than machine language?" Frank Beckman, who was also present, recalled that Von Neumann dismissed the whole development as "but an application of the idea of Turing's 'short code'." Donald Gillies, one of Von Neumann's students at Princeton, and later a faculty member at the University of Illinois, recalled in the mid-1970s that the graduates students were being "used" to hand assemble programs into binary for their early machine (probably the IAS machine). He took time out to build an assembler, but when Von Neumann found it out he was very angry, saying (paraphrased), "It is a waste of a valuable scientific computing instrument to use it to do clerical work".

One last anecdote about Von Neumann's brilliant mathematical capabilities. The Von Neumann household in Princeton was open to many social activities and on one such occasion someone posed the "fly and the train" problem to Von Neumann. Quickly Von Neumann came up with the answer. Suspecting that he had seen through the problem to discover a simple solution, he was asked how he solved the problem. "Simple", he responded, "I summed the series!" [From Nick Metropolis]

The Institute of Electrical and Electronics Engineers (IEEE) continues to honor John Von Neumann through the presentation of an annual award in his name. The IEEE John Von Neumann Medal was established by the Board of Directors in 1990 and may be presented annually "for outstanding achievements in computer-related science and technology". The achievements may be theoretical, technological, or entrepreneurial, and need not have been made immediately prior to the date of the award.

Quotations

If people do not believe that mathematics is simple, it is only because they do not realize how complicated life is.

Anyone who considers arithmetical methods of producing random digits is, of course, in a state of sin.

New Words

Budapest	['bu:dəpest]	*n.*布达佩斯(匈牙利首都)
Hungary	['hʌŋgərɪ]	*n.*匈牙利
tutelage	['tju:təlɪdʒ]	*n.*教导；指导
publish	['pʌblɪʃ]	*vt.*公开；发表；出版；发行
Berlin	[bə: 'lin]	*n.*柏林(德国城市)
Zurich	['zjʊərik]	*n.*苏黎世(瑞士城市)

diploma	[dɪ'pləʊmə]	*n.*文凭，毕业证书
algebra	['ældʒɪbrə]	*n.*代数学
unrest	[ʌn'rest]	*n.*不安的状态，动荡的局面
appoint	[ə'pɔɪnt]	*vt.*任命；委派
retain	[rɪ'teɪn]	*vt.*保持，保留
remainder	[rɪ'meɪndə]	*n.*剩余物
instigation	[ˌɪnstɪ'geɪʃn]	*n.*鼓动，教唆，煽动
sponsorship	['spɒnsəʃɪp]	*n.*发起，主办；倡议
clearance	['klɪərəns]	*n.*证明书，过失、可靠或称职的官方证明，批准证
anecdote	['ænɪkdəʊt]	*n.*趣闻；逸事
immigration	[ˌɪmɪ'greɪʃn]	*n.*移民总称；移居；移民入境
inconsistency	[ˌɪnkən'sɪstənsɪ]	*n.*矛盾
dissertation	[ˌdɪsə'teɪʃn]	*n.*(学位)论文
peer	[pɪə]	*n.*同等的人；同辈；同龄人
perceive	[pə'si:v]	*vt.*察觉；理解；发觉；意识到
expertise	[ˌekspɜ: 'ti:z]	*n.*专门技能；专门知识；专门技术
hydrodynamics	[ˌhaɪdrəʊdaɪ'næmɪks]	*n.*流体力学，水动力学
ballistics	[bə'lɪstɪks]	*n.*弹道学，发射学
meteorology	[ˌmi:tɪə'rɒlədʒɪ]	*n.*气象学；气象状态
statistics	[stə'tɪstɪks]	*n.*统计学；统计表
electromechanical	[ɪˌlektrəʊmɪ'kænɪkəl]	*adj.* [机]电动机械的，机电的，电机的
relay	['ri:leɪ]	*n.*继电器
conduit	['kɒndjuɪt]	*n.*管道，导管
capability	[ˌkeɪpə'bɪlətɪ]	*n.*能力；才能；才干
synergism	['sɪnədʒɪzəm]	*n.*合作
obituary	[ə'bɪtʃuərɪ]	*n.*讣告
multitudinousness	[ˌmʌltɪ'tjui:dɪnəsnɪs]	*n.*非常多；众多的
infrastructure	['ɪnfrəstrʌktʃə]	*n.*基本设施
parallelism	['pærəlelɪzəm]	*n.*平行，对应，类似
author	['ɔ:θə)]	*vt.*写作，创作 *n.*作家，创造者
retrospective	[ˌretrə'spektɪv]	*adj.*回顾的
entrepreneurial	[ˌɒntrəprə'nɜ:rɪəl]	*adj.*企业家的，创业者的

Phrases

child prodigy	神童
Chemical Engineering	化学工程
doctoral degree	博士学位
set theory	集合论

quantum mechanics　　量子力学
Princeton University　　普林斯顿大学
the US Constitution　　美国宪法
applied mathematics　　应用数学
game theory　　博弈论，对策论
be shielded from　　与……隔离
hydrogen bomb　　氢弹

Word Building

学习英语的关键与难点之一就是记忆单词。机械地逐一记忆单词会花费大量的时间。而且，在计算机行业中不断涌现出一些新构造出来的词，如 unformat，undelete，reset，uninstall 等，在字典中往往查不到这些单词。因此，必须学会科学地记忆单词和识别新词。

其实，英语单词有其内在的结构规律，这就是构词法。掌握了构词法，就可达到举一反三、见词识义的学习效果。因此，掌握构词法是记忆英语单词的捷径。

常用的构词法有合成、转化及派生 3 种。下面主要介绍合成法（compounding）。

由两个和两个以上的词合成一个新词的构词方法就叫"合成法"。用合成法构成的词叫作"复合词"。复合词可以有以下 3 种书写形式：连起来写，例如，网络 network；分开写，例如，汽车站 bus stop；用连字符连在一起，例如，内置的 built-in。由 3 个以上单词构成一个复合词时常采用第 3 种形式，例如，现代的 up-to-date，一对一的 one-to-one，容易使用的 easy-to-use。合成词的前一个词常用来说明后一个词，例如，主板 motherboard，子板 daughterboard。绝大部分合成词的词性由最后一个单词决定，但也有例外，例如，up-to-date 就是形容词。

合成词的构成方法如下，以合成名词 (compound noun)为例：

1. 名词+名词

wave + length —— wavelength 波长
band + width —— bandwidth 带宽，频带宽度
bench + mark —— benchmark 基准测试
bold + face —— boldface 粗体
clip + board —— clipboard 剪贴板
chip + set —— chipset 芯片组
copy + right —— copyright 版权
data + base —— database 数据库
finger + tip —— fingertip 手指尖
firm + ware —— firmware 固件
lap + top —— laptop 膝盖，膝面，膝上
screw + driver —— screwdriver 螺丝刀
snail + mail —— snailmail 慢邮件（平信）
spread + sheet —— spreadsheet 电子表格；数据表
web + site —— website 网站

2. 名词+动名词

machine + building —— machine building 机器制造

book + learning —— book learning 书本知识

hand + writing —— handwriting 手写，手稿

3. 动名词+名词

waiting + room —— waiting room 候车室

building + material —— building material 建筑材料

swimming + pool —— swimming pool 游泳池

4. 形容词+名词

short + hand —— shorthand 速记

hard+ ware —— hardware 硬件

soft + ware —— software 软件

lower + case —— lowercase 下档；小写字母

upper + case —— uppercase 上档；大写字母

broad + band —— broadband 宽波段

fresh + man —— freshman 大（中）学一年级学生

hard + copy —— hard copy 硬拷贝

5. 动词+名词

pick + pocket —— pickpocket 小偷

break + water —— breakwater 防波堤

6. 副词+动词

in + put —— input 输入

out + put —— output 输出，产量

out + come —— outcome 结果

7. 动词+副词

feed + back —— feedback 反馈

get + together —— get-together 联欢会

stand + still —— standstill 停顿

Career Training

招 聘 启 事

发布招聘启事是企业招聘人才的常用方法之一。它具有受众广泛、快捷方便及成本低廉的特点。尤其是通过互联网发布招聘启事，优点更为明显，这也是 IT 企业最热衷的方法。

招聘启事通常具有简洁明了及准确严谨的文风。因此，表格及规格化的表达比较常用，以实现有条理、有层次的表达。其开始往往是职位描述，然后是对应聘者的具体要求。一个规范的招聘启事会明确表示工作的行业、性质、是否需要经常出差及工作地点等信息。

在 IT 企业的招聘启事中，往往会使用许多专业词汇，特别是缩略语（如 PIX、VPNs、

TCP/IP、IDS 及 CCIE）。另外，在 IT 企业的招聘启事中，往往还会使用一些特殊的表达方式，如 4+ years experience 表示“四年以上经验”。

在阅读招聘启事时，应特别重视副词和形容词，如 Strong 及 In-depth。

Network Engineer

- Oversee all aspects of Network and Information Security.
- Manage the design, implementation, and audit of various security controls, as well as facilitate the policy creation and revision process.
- Ensure conformance with best practices across the enterprise.
- Recommend, configure, and deploy commercial or open source tools.
- Recommend and configure commercial firewalls, routers, switches.
- Work with leads to design, implement, manage n-tier and multi-segmented networks.
- Act in rotation as part of an on-call network operations team.

When submitting resume, please include salary history.

Required Skills:

Qualifications include:

- 4+ years experience in network, internetwork, security, and system administrations.
- Strong knowledge of Cisco PIX, switches, routers and load balancers, packet sniffers, remote access methodologies, and VPNs.
- Strong network performance tuning, and issue resolution skills.
- Configuration management methodologies.
- Understanding of TCP/IP, Solaris and Linux, network/system intrusion techniques.
- In-depth understanding of software security systems, including cryptography, certificate authorities, etc.
- Database security experience (roles, etc) as a plus.
- Network security knowledge (network based IDS, Firewalls, VPNs, etc).
- Host based security knowledge (patching, hardening).
- BS in Computer Science or equivalent work experience.
- Applicable industry certification desired (CCIE preferred).

Industry:	Financial
Emp. Type:	Full Time
Travel:	No Travel
Location:	Irvine, CA
Overtime Pay:	None
Date Posted:	11/23/2020

软件水平考试试题解析

【真题再现】

从供选择的答案中选出应填入下列英语文句中____内的正确答案，把编号写在答卷的对应栏内。

Here is a useful procedure for choosing a program:

1. Study the features of all the programs you might choose __A__. Decide which features you need, which you would __B__, and which you do not need.

2. Eliminate the programs that clearly do not __C__ your needs.

3. Consider how the remaining programs perform the functions you will use most often. This can affect a program's usability more than all the "nice" features that you will __D__ need.

4. Study the remaining programs carefully with __E__ experience if you can get it and decide which one is best for you.

供选择的答案：

A： ①for ②on ③in ④from ⑤choose ⑥like

B，C： ①meet ②require ③help ④give ⑤choose ⑥like

D，E： ①often ②seldom ③always ④rich ⑤hands-on ⑥little

【答案】 A：④ B：⑥ C：① D：② E：⑤

【试题解析】

A：choose from…的意思是“从……中选出”。本句的意思是“研究备选程序的特性”。故选④。

B：meet 的意思是“满足”，require 的意思是“需要”，help 的意思是“帮助”，give 的意思是“给予”，choose 的意思是“选择”，like 的意思是“喜欢”。would like 的意思是“想要，喜欢”。which you would like 意思是“你想要的那些性能”。故选⑥。

C：meet one's needs 是一个固定词组，意思是“满足某人的需要”。故选①。

D：根据句子结构，应该填一个副词。often，seldom 和 always 都是副词，表示频率。always 的频率最高，意思是“总是”。请看如下例句：

You can always find him working on his computer.

你总能看见他在计算机上工作。

often 的意思是“经常，通常”。

seldom 的意思是“不经常；很少或难得”。请看如下例句：

Do you often play computer games? Seldom.

你经常玩电脑游戏吗？不，很少玩。

根据句意，此处应该填与 most often 相反的词。故选②。

E：根据句子结构，应该填一个形容词。rich 的意思是“丰富的”。hands-on 的意思是“亲身实践的”，little 的意思是“很少的，几乎没有的”。结合句意，故选⑤。

【参考译文】

下面是选择程序的实用步骤：

1. 研究备选程序的特性。决定哪些特性是所需的，哪些特性是想要的，哪些特性是不需要的。

2. 去掉那些明显不符合需求的程序。

3. 对于剩下的程序，考虑如何执行经常使用的功能。这会影响程序的可用性，而不是很少需要的“好”特性。

4. 如果有实际经验的话，仔细研究剩下的程序，然后选择最适合的程序。

Exercises

[Ex. 1] 根据 Text A 回答以下问题。

1) What is a computer?

2) What does a modern computer consists of conventionally?

3) What does the term hardware cover?

4) How many main components does a general purpose computer have? What are they?

5) What are some examples of input devices mentioned in the passage?

6) What are some examples of output devices mentioned in the passage?

7) What has happened to CPUs since the mid-1970s?

8) What do logic operations involve?

9) What are registers used for?

10) What are the two principal varieties of computer main memory?

[Ex. 2] 根据 Text B 回答以下问题。

1) Where was Von Neumann born?

2) When did Von Neumann enter the University of Budapest? What did he study there?

3) When did Von Neumann return to his first love of mathematics?

4) When was Von Neumann appointed to be one of the original six Professors of Mathematics?

5) How did Von Neumann's interest in computers differ from that of his peers?

6) During the latter years of World War II what part was Von Neumann playing?

7) What did postwar Von Neumann concentrate on?

8) How were the basic elements of the stored program concept introduced to the industry?

9) In the 1950s what was Von Neumann employed as?

10) When was the IEEE John Von Neumann Medal established? And by whom?

[Ex. 3] 把下列句子翻译为中文。

1) These newly-designed devices will appear on the exhibition next month.

2) Each of the components is useful in its degree.

3) The markings are so blurred that it is difficult to identify.

4) An external modem is a stand-alone modem that is connected via cable to a computer's serial port.

5) All information, from train schedules to discount-price goods, will be as close as the press of a key.

6) CPU is the brain of a computer.

7) A desktop computer is small enough to fit conveniently on the surface of a business desk.

8) A mouse is a relative pointing device because there are no defined limits to the mouse's movement and because its placement on a surface does not map directly to a specific screen location.

9) Without communications software, however, modems cannot perform any useful work.

10) A keyboard on a computer is almost identical to a keyboard on a typewriter, except it has extra keys.

[Ex. 4] 软件水平考试真题自测。

从供选择的答案中选出应填入下列英语文句中____内的正确答案，把编号写在答卷的对应栏内。

Perhaps you have been asking the questions: Is everyone moving to __A__? If I don't step up to the __B__ user interface, will I be left all along at the cold __C__ command line?

Don't throw away your DOS __D__. No one is __E__ DOS's death knell yet. DOS and its applications people have been using for years are not going to stop working overnight.

供选择的答案：

A～E：①DOS ②UNIX ③Windows ④Windows NT ⑤graphical
⑥software ⑦replacing ⑧sounding ⑨text ⑩hardware

[Ex. 5] 听短文填空。

Exercises 1

Picking a new CPU can be a daunting ____1____. Whether you're a DIY aficionado or are deciding which CPU to ____2____ in a new desktop, knowing what tradeoffs to make ____3____ price and performance is always ____4____. Do you live in the low-CPU-usage world of word ____5____, spreadsheets, and Web browsers, or are you a power-hungry 3D gamer, ____6____ recorder, or video-editing expert who ____7____ all the horsepower you can get? 3D games and ____8____ encoding, in particular, really pound on the CPU. What if you use all of these applications and your budget isn't unlimited? That's where knowing how to get the ____9____ CPU at the right ____10____ really comes in handy.

Reference Translation

计算机硬件

计算机是指可以通过计算机编程自动执行算术或逻辑运算的机器。现代计算机能够遵循被称为程序的通用操作集。这些程序使计算机能够执行极其广泛的任务。"完整"的计算机包括硬件、操作系统和外围设备，它们可以满足计算机系统所需的"全部"操作。该计算机可以被称为一个计算机系统。该术语也可以用来指连接在一起的一组计算机，特别是计算机网络或计算机集群。

计算机用作各种工业和消费设备的控制系统。这些设备包括简单的专用设备（如微波炉和遥控器）、工业设备（如工业机器人和计算机辅助设计设备），以及通用设备（如个人计算机）和移动设备（如智能手机）。互联网在计算机上运行，它连接了数以亿计的计算机及其用户。

通常，现代计算机至少包括一个处理组件（通常是中央处理器，CPU）和某种形式的存储器。处理组件执行算术和逻辑运算，并且排序和控制单元可以响应存储的信息来改变操作的顺序。外围设备包括输入设备（键盘、鼠标等）、输出设备（监视器屏幕、打印机等）。外围设备允许从外部源获得信息，并且使操作结果得以保存和访问。

术语"硬件"涵盖计算机中全部有形的物理部件。电路、计算机芯片、图形卡、声卡、存储器（随机存取存储器，RAM）、主板、显示器、电源、电缆、键盘、打印机和鼠标输入设备都是硬件。

通用计算机具有四种主要部件：算术逻辑部件（ALU）、控制单元、存储器以及输入输出设备（统称为 I/O）。这些部件通过总线互连，总线通常由电线组成。这些部件内部都有数千到数万亿的小电路，可以通过电子开关关闭或打开。每个电路表示一位（二进制数字）信息，当电路接通时它表示"1"，而当电路断开时则表示"0"。这些电路被布置在逻辑门中，以便一个或多个电路可以控制一个或多个其他电路的状态。

1. 输入设备

输入设备将未处理的数据发送到计算机时，数据将被处理并发送到输出设备。输入设备既可以手动操作，也可以自动操作。处理行为主要由中央处理器（CPU）控制。输入设备有计算机键盘、数码相机、数码摄像机、图形输入板、图像扫描仪、传声器、鼠标、触摸屏。

2. 输出设备

计算机提供输出的设备称为输出设备。输出设备有计算机监视器、打印机、PC 扬声器、声卡、视频卡。

3. 控制单元

控制单元（通常称为控制系统或中央控制器）管理计算机的各种组件；它读取并解释（解码）程序指令，将它们转换为激活计算机其他部件的控制信号。高级计算机中的控制系统可以改变某些指令的执行顺序以提高性能。

所有 CPU 共有的关键组件是程序计数器，它是一个特殊的存储单元（寄存器），用于跟踪存储器中下一条指令的读取位置。

控制系统的功能如下。请注意，这是一个简化的描述，其中一些步骤可以同时执行或以不同的顺序执行，具体取决于 CPU 的类型：

（1）从程序计数器指示的单元中读取下一条指令的代码。

（2）将指令的数字代码解码为其他各系统的一组命令或信号。

（3）递增程序计数器的值，使其指向下一条指令。

（4）从存储器单元中（或者可能来自输入设备）读取指令所需的任何数据。所需数据的位置通常存储在指令代码中。

（5）向 ALU 或寄存器提供必要的数据。

（6）如果指令需要 ALU 或专用硬件来完成，则指示硬件执行所请求的操作。

（7）将结果从 ALU 写回存储器位置、寄存器或输出设备。

（8）回到第一步。

4. 中央处理器（CPU）

控制单元、ALU 和寄存器统称为中央处理器（CPU）。早期的 CPU 由许多独立的组件组成，但自 20 世纪 70 年代中期以来，CPU 通常构建在称为微处理器的单个集成电路上。

5. 算术逻辑部件（ALU）

ALU 能够执行两类操作：算术和逻辑。特定 ALU 支持的算术运算集可以仅限于加法和减法，也可以包括乘法、除法、三角函数（例如正弦、余弦等）及平方根。有些运算只能使用整数，而有些则使用浮点来表示实数（尽管精度有限）。然而，任何能够执行最简单操作的计算机都可以编程，只要将复杂的操作分解为可以执行的简单步骤。因此，可以对任何计算机进行编程以执行任何算术运算，但如果 ALU 不直接支持这些操作则需要更多时间。ALU 还可以比较数字并返回布尔值（真或假），这取决于比较结果是等于、大于还是小于。逻辑运算涉及布尔逻辑：AND、OR、XOR 和 NOT。它们对于创建复杂的条件语句和处理布尔逻辑非常有用。

超标量计算机可能包含多个 ALU，允许它们同时处理多个指令。具有单指令多数据（SIMD）和多指令多数据（MIMD）功能的图形处理器和计算机通常包含可以对矢量和矩阵执行算术的 ALU。

6. 存储器

计算机的存储器可被视为能够放置或读取数字的单元组。每个单元格都有一个编号——“地址”，可以存储一个数字。存储在存储器中的信息实际上可以代表任何东西，字母、数字甚至计算机指令都可以同样轻松地放入存储器中。由于 CPU 不区分信息的类型，因而软件负责解释存储器中信息的意义，但对存储器来说，这些信息只不过是一系列数字。

在几乎所有现代计算机中，每个存储器单元都被设置为以 8 位（即字节）为一组来存储二进制数。每个字节能够代表 256 个不同的数字（$2^8 = 256$）：从 0～255，或-128～+127。为了存储更大的数字，可以使用几个连续的字节（通常是 2 个、4 个或 8 个）。当需要存储负数时，通常以二进制补码的形式来存储。任何类型的信息只要能表示为数字，计算机就能将其存储在存储器中。

CPU 包含一组称为寄存器的特殊存储器单元，读取和写入寄存器的速度比主存储器更快。根据 CPU 的类型，通常有 2～100 个寄存器。寄存器用于最常用的数据项，以免每次需要数据时都必须访问主存储器。因为数据会不断被处理，减少对主存储器的访问（与 ALU 和控制单元相比访问主存储器的速度通常很慢），大大提高了计算机的速度。

计算机主存储器有两种主要类型：

- 随机存取存储器，即 RAM。
- 只读存储器，即 ROM。

CPU 可以在任何时候读取和写入 RAM，而 ROM 因为预装了永不改变的数据和软件，所以 CPU 只能读取它。ROM 通常用于存储计算机的初始启动指令。一般情况下，当关闭计算机电源时，RAM 的内容会被删除，但 ROM 中的数据会无限期保留。在个人计算机中，ROM 包含一个称为 BIOS 的专用程序，它可以在计算机打开或重置时协调将计算机操作系统从硬盘驱动器加载到 RAM 中。在没有磁盘驱动器的嵌入式计算机中，所有需要的软件都可以存储在 ROM 中。存储在 ROM 中的软件通常称为固件，因为它在概念上更像是硬件而不是软件。闪存模糊了 ROM 和 RAM 之间的区别，因为它在计算机关闭时保留其数据但也可以重写。

7. 多重处理

一些计算机被设计为具有多重处理配置，它能将其工作分布在多个 CPU 上，这种技术曾经只用于超级计算机、大型计算机和服务器等功能强大的机器。现在多重处理器和多核（单个集成电路上的多个 CPU）在个人计算机和膝上计算机中得到广泛使用。

特别是超级计算机，通常具有独特的架构，与基本存储程序和通用计算机显然不同。它们通常具有数千个 CPU、定制的高速互连和专用计算硬件。这种设计往往用于专门的任务。超级计算机通常用于大规模模拟、图形渲染和加密应用程序。

Unit 2

Text A

Software

Software is a general term for the various kinds of programs used to operate computers and related devices. (The term hardware describes the physical aspects of computers and related devices.)

Text A
Software

Software can be thought of as the variable part of a computer and hardware the invariable part. Software is often divided into application software (programs that do work users are directly interested in) and system software (which includes operating systems and any program that supports application software).[1] The term middleware is sometimes used to describe programming that mediates between application and system software or between two different kinds of application software (for example, sending a remote work request from an application in a computer that has one kind of operating system to an application in a computer with a different operating system).

An additional and difficult-to-classify category of software is the utility, which is a small useful program with limited capability.[2] Some utilities come with operating systems. Like applications, utilities tend to be separately installable and capable of being used independently from the rest of the operating system.

Applets are small applications that sometimes come with the operating system as "accessories". They can also be created independently using the Java or other programming languages.

Software can be purchased or acquired as shareware (usually intended for sale after a trial period), liteware (shareware with some capabilities disabled), freeware (free software but with copyright restrictions), public domain software (free with no restrictions), and open source (software where the source code is furnished and users agree not to limit the distribution of improvements).

Software was often packaged on CD-ROMs. Today, much purchased software, shareware, and freeware is downloaded over the Internet. A new trend is software that is made available for use at another site known as an application service provider.

1. Operating System

An operating system (sometimes abbreviated as "OS") is the program that, after being initially loaded into the computer by a boot program, manages all the other programs in a computer. The other programs are called applications or application programs. The application programs make use of the operating system by making requests for services through a defined application program interface (API). In addition, users can interact directly with the operating system through a user interface such as a command language or a graphical user interface (GUI).

An operating system performs these services for applications:

- In a multitasking operating system where multiple programs can be running at the same time, the operating system determines which applications should run in what order and how much time should be allowed for each application before giving another application a turn.[3]
- It manages the sharing of internal memory among multiple applications.
- It handles input and output to and from attached hardware devices, such as hard disks, printers, and dial-up ports.
- It sends messages to each application or interactive user (or to a system operator) about the status of operation and any errors that may have occurred.
- It can offload the management of what are called batch jobs (for example, printing) so that the initiating application is freed from this work.
- On computers that can provide parallel processing, an operating system can manage how to divide the program so that it runs on more than one processor at a time.

2. Application software

Application software is all the computer software that causes a computer to perform useful tasks beyond the running of the computer itself. A specific instance of such software is called a software application, application program, application or APP.

The term is used to contrast such software with system software, which manages and integrates a computer's capabilities but does not directly perform tasks that benefit the user.[4] The system software serves the application, which in turn serves the user.

Examples include accounting software, enterprise software, graphics software, media players, and office suites. Many application programs deal principally with documents. Applications may be bundled with the computer and its system software or published separately, and can be coded as university projects.

Application software applies the power of a particular computing platform or system software to a particular purpose.

Some applications are available in versions for several different platforms; others have narrower requirements and are thus called, for example, a Geography application for Windows, an Android application for education, or Linux gaming. Sometimes a new and popular application arises which only runs on one platform, increasing the desirability of that platform. This is called a killer application.

There are many different ways to divide up different types of application software, and several are explained here.

Since the development and near-universal adoption of the Web, an important distinction that has emerged has been between Web applications — written with HTML, JavaScript and other Web-native technologies and typically requiring one to be online and running a Web browser, and the more traditional native applications written in whatever languages are available for one's particular type of computer. There has been contentious debate in the computing community regarding Web applications replacing native applications for many purposes, especially on mobile devices such as smart phones and tablets. Web APPs have indeed greatly increased in popularity for some uses, but the advantages of native applications make them unlikely to disappear soon. Furthermore, the two can be complementary, and even integrated.

Application software can also be seen as being either horizontal or vertical. Horizontal applications are more popular and widespread, for example, word processors or databases. Vertical applications are niche products, designed for a particular type of industry or business, or department within an organization. Integrated suites of software will try to handle every specific aspect possible of, for example, manufacturing or banking systems, or accounting, or customer service.

There are many types of application software:

- An application suite consists of multiple applications bundled together. They usually have related functions, features and user interfaces, and may be able to interact with each other, e.g. open each other's files. Business applications often come in suites, e.g. Microsoft Office, LibreOffice and iWork, which bundle together a word processor, a spreadsheet, etc.; but suites exist for other purposes, e.g. graphics or music.[5]
- Enterprise software addresses the needs of an entire organization's processes and data flow, across most all departments, often in a large distributed environment. Examples include financial systems, customer relationship management (CRM) systems and supply chain management software. Departmental Software is a sub-type of enterprise software with a focus on smaller organizations and/or groups within a large organization. Examples include travel expense management and IT Helpdesk.
- Enterprise infrastructure software provides common capabilities needed to support enterprise software systems. Examples include databases, E-mail servers, and systems for managing networks and security.
- Information worker software lets users create and manage information, often for individual projects within a department, in contrast to enterprise management. Examples include time management, resource management, documentation tools, analytical and collaborativetools. Word processors, spreadsheets, E-mail and blog clients, personal information system, and individual media editors may aid in multiple information worker tasks.
- Content access software is used primarily to access content without editing, but may include software that allows for content editing. Such software addresses the needs of individuals and groups to consume digital entertainment and published digital content. Examples include media players, Web browsers, and help browsers.
- Educational software is related to content access software, but has the content and/or features adapted for use by educators or students. For example, it may deliver evaluations (tests), track progress through material.
- Simulation software simulates physical or abstract systems for research, training or entertainment purposes.

- Media development software generates print and electronic media for others to consume, most often in a commercial or educational setting. This includes graphic-art software, desktop publishing software, multimedia development software, HTML editors, digital-animation editors, digital audio and video composition, and many others.
- Product engineering software is used in developing hardware and software products. This includes computer aided design (CAD), computer aided engineering (CAE), computer language editing and compiling tools, integrated development environments, and application program interfaces.

Applications can also be classified by computing platform such as a particular operating system, delivery network such as in cloud computing and Web 2.0 applications, or delivery devices such as mobile APPs for mobile devices.

The operating system itself can be considered application software when performing simple calculating, measuring, rendering, and word processing tasks. This does not include application software bundled within operating systems such as a software calculator or text editor.

3. Middleware

In the computer industry, middleware is a general term for any programming that serves to "glue together" or mediate between two separate and often already existing programs. A common application of middleware is to allow programs written for access to a particular database to access other databases.

Typically, middleware programs provide messaging services so that different applications can communicate. The systematic tying together of disparate applications, often through the use of middleware, is known as enterprise application integration (EAI).

4. Firmware

In electronic systems and computing, firmware is the combination of persistent memory and program code and data stored in it. Typical examples of devices containing firmware are embedded systems (such as traffic lights, consumer appliances, and digital watches), computers, computer peripherals, mobile phones, and digital cameras. The firmware contained in these devices provides the control program for the device. Firmware is held in nonvolatile memory devices such as ROM, EPROM, or flash memory. Changing the firmware of a device may rarely or never be done during its economic lifetime; some firmware memory devices are permanently installed and cannot be changed after manufacture. Common reasons for updating firmware include fixing bugs or adding features to the device. This may require physically changing ROM integrated circuits, or reprogramming flash memory with a special procedure. Firmware such as the ROM BIOS of a personal computer may contain only elementary basic functions of a device and may only provide services to higher-level software. Firmware such as the program of an embedded system may be the only program that will run on the system and provide all of its functions.

software ['sɒftweə] *n.*软件

invariable	[ɪn'veəriəbl]	*adj*.不变的，常量的
middleware	['mɪdlweə]	*n*.中间件
category	['kætəgərɪ]	*n*.种类，类别
utility	[ju: 'tɪlətɪ]	*n*.实用（程序），效用，有用
independently	[ˌɪndɪ'pendəntlɪ]	*adv*.独立地
applet	['æplət]	*n*.小应用程序
shareware	['ʃeəweə]	*n*.共享软件
liteware	[laɪtweə]	*n*.精简版软件
freeware	['fri:weə]	*n*.免费软件
restriction	[rɪ'strɪkʃn]	*n*.限制，约束
furnish	['fɜ:nɪʃ]	*vt*.供应，提供
package	['pækɪdʒ]	*n*.包裹，包
download	[ˌdaʊn'ləʊd]	*v*.下载
initially	[ɪ'nɪʃəlɪ]	*adv*.最初，开头
boot	[bu:t]	*v*.引导，导入
application	[ˌæplɪ'keɪʃn]	*n*.应用，应用程序，应用软件
interact	[ˌɪntər'ækt]	*vi*.互相作用，互相影响
dial-up	['daɪəl-ʌp]	*v*.拨号（上网）
port	[pɔ:t]	*n*.端口
interactive	[ˌɪntər'æktɪv]	*adj*.交互式的
batch	[bætʃ]	*n*.批处理
free	[fri:]	*adj*.自由的，空闲的 *vt*.释放
parallel	['pærəlel]	*adj*.并行的，相同的
instance	['ɪnstəns]	*n*.实例，建议，要求，情况，场合 *vt*.举……为例，获得例证
integrate	['ɪntɪgreɪt]	*vt*.使成整体，使一体化
serve	[sɜ:v]	*v*.服务，供应，适合
graphic	['græfɪk]	*n*.图形
deal	[di:l]	*vi*.处理，应付
bundle	['bʌndl]	*n*.捆，束，包 *v*.捆扎
separately	['seprətlɪ]	*adv*.个别地，分离地
platform	['plætfɔ:m]	*n*.平台
available	[ə'veɪləbl]	*adj*.可用到的，可利用的，有用的
narrow	['nærəʊ]	*adj*.狭窄的，精密的，严密的，有限的
Android	['ændrɔɪd]	*n*.安卓
desirability	[dɪˌzaɪərə'bɪlətɪ]	*n*.愿望，希求
Web	[web]	*n*.万维网

emerge	[ɪ'mɜ:dʒ]	*vi.*显现，浮现，暴露，形成
traditional	[trə'dɪʃənl]	*adj.*传统的，惯例的
contentious	[kən'tenʃəs]	*adj.*争论的，有异议的
community	[kə'mju:nətɪ]	*n.*社区，团体
popularity	[ˌpɒpju'lærətɪ]	*n.*普及，流行
niche	[nɪtʃ]	*n.*细分市场，商机
interface	['ɪntəfeɪs]	*n.*接口，界面
spreadsheet	['spredʃi:t]	*n.*电子制表软件，电子数据表
enterprise	['entəpraɪz]	*n.*企业，事业
distribute	[dɪ'strɪbju:t]	*vt.*分发，分布
E-mail	['i:meɪl]	*n.*电子邮件
security	[sɪ'kjʊərətɪ]	*n.*安全
analytical	[ˌænə'lɪtɪkl]	*adj.*分析的，解析的
collaborative	[kə'læbərətɪv]	*adj.*合作的，协作的
aid	[eɪd]	*n.*& *vt.*辅助，帮助
progress	['prəʊgres]	*n.*进程，发展
simulation	[ˌsɪmju'leɪʃn]	*n.*仿真，模拟
simulate	['sɪmjuleɪt]	*vt.*模拟，模仿
generate	['dʒenəreɪt]	*vt.*产生，发生
audio	['ɔ:dɪəʊ]	*adj.*音频的，声频的，声音的
compile	[kəm'paɪl]	*vt.*编译，编辑，汇编
delivery	[dɪ'lɪvərɪ]	*n.*递送，交付；发送，传输
systematic	[ˌsɪstə'mætɪk]	*adj.*系统的，体系的
firmware	['fɜ:mweə]	*n.*固件，韧件（软件和硬件相结合）
nonvolatile	['nɒn'vɒlətaɪl]	*adj.*非易失性的
permanently	['pɜ:mənəntlɪ]	*adv.*永存地，不变地
update	[ʌp'deɪt]	*vt.*更新，使现代化
	['ʌpdeɪt]	*n.*更新
bug	[bʌg]	*n.*程序缺陷
reprogram	[rɪ'prəʊgræm]	*v.*重新编程，程序重调
elementary	[ˌelɪ'mentrɪ]	*adj.*初步的，基本的
provide	[prə'vaɪd]	*v.*供应，供给，准备

Phrases

be thought of	被认为
be divided into	被分为
system software	系统软件
application software	应用软件

programming language	编程语言
trial period	试用期
open source	开源
application service provider	应用服务提供商
application program	应用程序
hard disk	硬盘
parallel processing	并行处理
media player	媒体播放器
computing platform	计算平台
killer application	杀手级应用
divide up	分割
smart phone	智能电话
data flow	数据流
focus on	致力于，使聚焦于
integrated development environments	集成开发环境
glue together	胶合，黏合
embedded system	嵌入式系统
digital camera	数码相机

Abbreviations

CD-ROM (Compact Disc，Read-Only-Memory)	只读光盘
API (Application Program Interface)	应用程序接口
GUI (Graphical User Interface)	图形用户界面，图形用户接口
HTML (HyperText Markup Language)	超文本标识语言
CRM (Customer Relationship Management)	客户关系管理
IT (Information Technology)	信息技术
CAD (Computer Aided Design)	计算机辅助设计
CAE (Computer Aided Engineering)	计算机辅助工程
EAI (Enterprise Application Integration)	企业应用集成
ROM (Read Only Memory)	只读存储器
EPROM (Erasable Programmable Read-Only Memory)	可擦可编程只读存储器
BIOS (Basic Input Output System)	基本输入输出系统

Notes

[1] Software is often divided into application software (programs that do work users are directly interested in) and system software (which includes operating systems and any program that supports application software).

本句中，programs that do work users are directly interested in 是对application software 的补充说明，that do work users are directly interested in 是一个定语从句，修饰和限定 programs。

在该定语从句中，users are directly interested in 也是一个定语从句，修饰和限定 work。which includes operating systems and any program that supports application software 是一个定语从句，对system software 进行解释说明。在该定语从句中 that supports application software 也是一个定语从句，修饰和限定 any program。

[2] An additional and difficult-to-classify category of software is the utility, which is a small useful program with limited capability.

本句中，which is a small useful program with limited capability 是一个非限定性定语从句，对 the utility 进行解释说明。

[3] In a multitasking operating system where multiple programs can be running at the same time, the operating system determines which applications should run in what order and how much time should be allowed for each application before giving another application a turn.

本句中，where multiple programs can be running at the same time 是一个定语从句，修饰和限定 a multitasking operating system。which applications should run in what order and how much time should be allowed for each application before giving another application a turn 是宾语从句，作 determines 的宾语。

[4] The term is used to contrast such software with system software, which manages and integrates a computer's capabilities but does not directly perform tasks that benefit the user.

本句中，The term 指上一段提到的 Application software。which manages and integrates a computer's capabilities but does not directly perform tasks that benefit the user 是一个非限定性定语从句，对 system software 进行解释说明。which 指 system software。that benefit the user 是一个定语从句，修饰和限定 tasks。

[5] Business applications often come in suites, e.g. Microsoft Office, LibreOffice and iWork, which bundle together a word processor, a spreadsheet, etc.; but suites exist for other purposes, e.g. graphics or music.

本句中，come in suites 的意思是"做成套件，以套件形式出现"。"which bundle together a word processor, a spreadsheet, etc"是一个非限定性定语从句，对"Microsoft Office, LibreOffice and iWork"进行补充说明。

Text B

Software Development Process

Text B
Software Development
Process

A software development process, also known as a software development life-cycle (SDLC), is a structure imposed on the development of a software product. Similar terms include software life cycle and software process. It is often considered as a subset of systems development life cycle. There are several models for such processes, each describing approaches to a variety of tasks or activities that take place during the process. Some people consider as a life-cycle model as a more general term and a software development process as a more specific term. For example, there are many specific software development processes that 'fit' the spiral life-cycle model. ISO/IEC 12207 is an international standard for software life-cycle processes. It aims to be the standard that defines all the tasks required for developing and maintaining software.

1. Overview

The large and growing body of software development organizations implement process methodologies. Many of them are in the defense industry, which in the U.S. requires a rating based on "process models" to obtain contracts.

The international standard for describing the method of selecting, implementing and monitoring the life cycle for software is ISO/IEC 12207.

A decades-long goal has been to find repeatable, predictable processes that improve productivity and quality. Some try to systematize or formalize the seemingly unruly task of writing software. Others apply project management techniques to writing software. Without effective project management, software projects can easily be delivered late or over budget. With large numbers of software projects not meeting their expectations in terms of functionality, cost, or delivery schedule, it is effective project management that appears to be lacking.

Organizations may create a software engineering process group (SEPG), which is the focal point for process improvement. Composed of line practitioners who have varied skills, the group is at the center of the collaborative effort of everyone in the organization who is involved with software engineering process improvement.

2. Software development activities

2.1 Planning

Planning is an objective of each and every activity, where we want to discover things that belong to the project. An important task in creating a software program is extracting the requirements or requirements analysis. Customers typically have an abstract idea of what they want as an end result, but do not know what software should do. Skilled and experienced software engineers recognize incomplete, ambiguous, or even contradictory requirements at this point. Frequently demonstrating live code may help reduce the risk that the requirements are incorrect.

Once the general requirements are gathered from the client, an analysis of the scope of the development should be determined and clearly stated. This is often called a scope document.

Certain functionality may be out of scope of the project as a function of cost or as a result of unclear requirements at the start of development. If the development is done externally, this document can be considered a legal document so that if there are ever disputes, any ambiguity of what was promised to the client can be clarified.

2.2 Implementation, testing and documenting

Implementation is the part of the process where software engineers actually program the code for the project.

Software testing is an integral and important phase of the software development process. This part of the process ensures that defects are recognized as soon as possible.

Documenting the internal design of software for the purpose of future maintenance and enhancement is done throughout development. This may also include the writing of an API, be it external or internal. The software engineering process chosen by the developing team will determine how much internal documentation (if any) is necessary. Plan-driven models (e.g., Waterfall) generally produce more documentation than Agile models.

2.3 Deployment and maintenance

Deployment starts directly after the code is appropriately tested, approved for release, and sold or otherwise distributed into a production environment. This may involve installation, customization (such as by setting parameters to the customer's values), testing, and possibly an extended period of evaluation.

Software training and support is important, as software is only effective if it is used correctly.

Maintaining and enhancing software to cope with newly discovered faults or requirements can take substantial time and effort, as missed requirements may force redesign of the software.

3. Software development models

Several models exist to streamline the development process. Each one has its pros and cons, and it is up to the development team to adopt the most appropriate one for the project. Sometimes a combination of the models may be more suitable.

3.1 Waterfall model

The waterfall model (Figure 2-1) shows a process, where developers are to follow these phases in order:

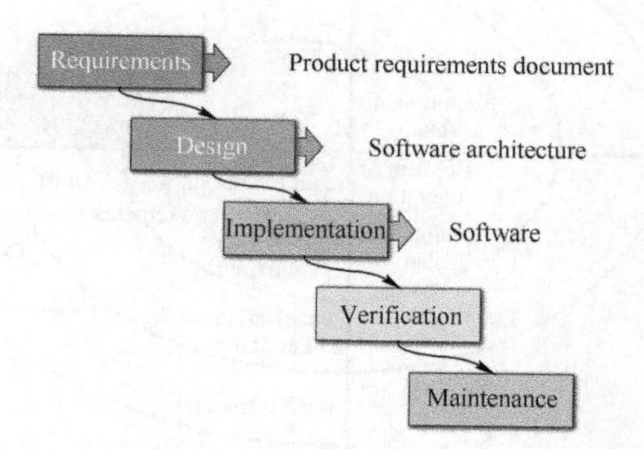

Figure 2-1 The activities of the software development process represented in the waterfall model

- Requirements specification (Requirements analysis).
- Software design.
- Implementation and Integration.
- Testing (or Validation).
- Deployment (or Installation).
- Maintenance.

In a strict waterfall model, after each phase is finished, it proceeds to the next one. Reviews may occur before moving to the next phase which allows for the possibility of changes (which may involve a formal change control process). Reviews may also be employed to ensure that the phase is indeed complete; the phase completion criteria are often referred to as a "gate" that the project must pass through to move to the next phase. Waterfall discourages revisiting and revising any prior phase once it's complete. This "inflexibility" in a pure Waterfall model has been a source of criticism by supporters of other more "flexible" models.

The waterfall model is also commonly taught with the mnemonic A Dance in the Dark Every Monday, representing Analysis, Design, Implementation, Testing, Documentation and Execution, and Maintenance.

3.2 Spiral model

The key characteristic of a spiral model is risk management at regular stages in the development cycle. In 1988, Barry Boehm published a formal software system development "spiral model", which combines some key aspect of the waterfall model and rapid prototyping methodologies, but provided emphasis in a key area many felt had been neglected by other methodologies: deliberate iterative risk analysis, particularly suited to large-scale complex systems.

The spiral model (Figure 2-2) is visualized as a process passing through some number of iterations, with the four quadrant diagram representative of the following activities:

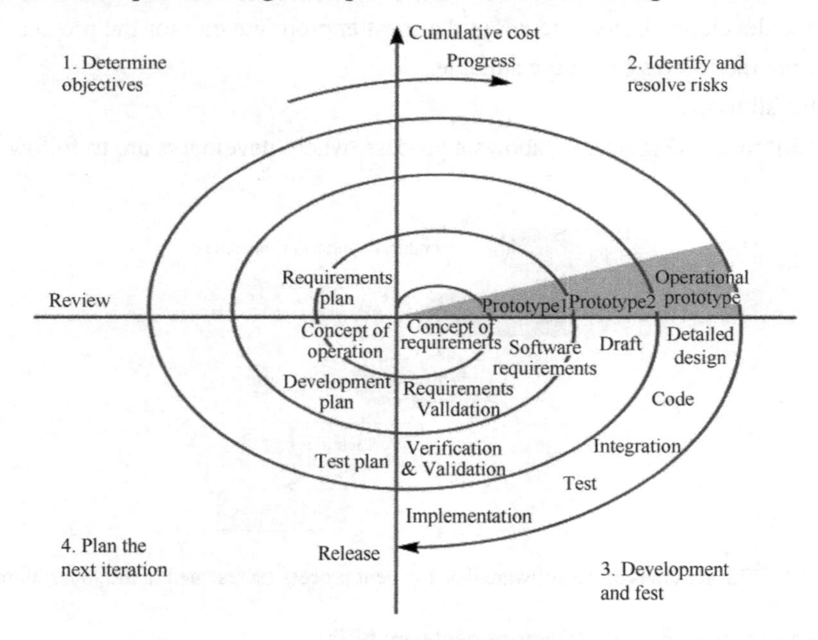

Figure 2-2 Spiral model

- Formulate plans: to identify software targets, implement the program, clarify the project development restrictions.
- Risk analysis: an analytical assessment of selected programs, to consider how to identify and eliminate risk.

- Implementation of the project: the implementation of software development and verification.

Risk-driven spiral model, emphasizing the conditions of options and constraints in order to support software reuse and software quality, can help as a special goal of integration into the product development. However, the spiral model has some restrictive conditions, as follows:

- The spiral model emphasizes risk analysis, and thus requires customers to accept this analysis and act on it. This requires both trust in the developer as well as the willingness to spend more to fix the issues, which is the reason why this model is often used for large-scale internal software development.
- If the implementation of risk analysis will greatly affect the profits of the project, the spiral model should not be used.
- Software developers have to actively look for possible risks, and analyze it accurately for the spiral model to work.

The first stage is to formulate a plan to achieve the objectives with these constraints, and then strive to find and remove all potential risks through careful analysis and, if necessary, by constructing a prototype. If some risks can not be ruled out, the customer has to decide whether to terminate the project or to ignore the risks and continue anyway. Finally, the results are evaluated and the design of the next phase begins.

3.3 Iterative and incremental development

The basic idea behind iterative development mode (Figure 2-3) is to develop a system through repeated cycles (iterative) and in smaller portions at a time (incremental), allowing software developers to take advantage of what was learned during development of earlier parts or versions of the system. Learning comes from both the development and use of the system, where possible key steps in the process start with a simple implementation of a subset of the software requirements and iteratively enhance the evolving versions until the full system is implemented. At each iteration, design modifications are made and new functional capabilities are added.

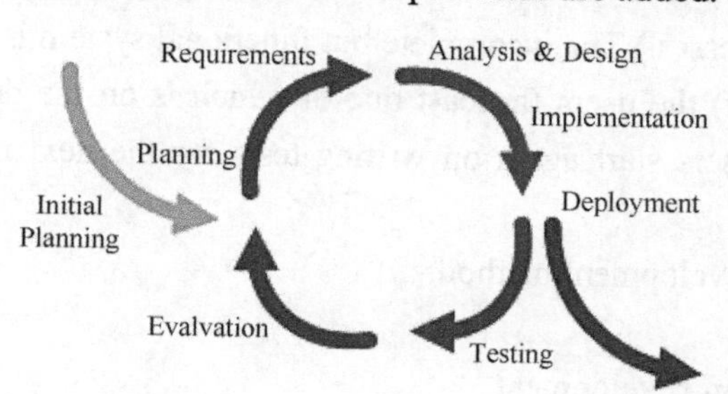

Figure 2-3 Iterative development model

The procedure itself consists of the initialization step, the iteration step, and the project control list. The initialization step creates a base version of the system. The goal for this initial implementation is to create a product to which the user can react. It should offer a sampling of the key aspects of the problem and provide a solution that is simple enough to understand and implement easily. To guide the iteration process, a project control list is created that contains a record of all tasks that need to be performed. It includes such items as new features to be implemented and areas of redesign of the existing solution. The control list is constantly being revised as a result of the analysis phase.

The iteration involves the redesign and implementation of iteration is to be simple, straightforward, and modular, supporting redesign at that stage or as a task added to the project control list. The level of design detail is not dictated by the iterative approach. In a light-weight iterative project the code may represent the major source of documentation of the system; however, in a critical iterative project a formal software design document may be used. The analysis of an iteration is based upon user feedback, and the program analysis facilities available. It involves analysis of the structure, modularity, usability, reliability, efficiency, and achievement of goals. The project control list is modified in light of the analysis results.

3.4 Agile software development

Agile software development uses iterative development as a basis but advocates a lighter and more people-centric viewpoint than traditional approaches. Agile processes fundamentally incorporate iteration and the continuous feedback that it provides to successively refine and deliver a software system.

There are many variations of agile processes:

- In extreme programming (XP), the phases are carried out in extremely small (or "continuous") steps compared to the older "batch" processes. The (intentionally incomplete) first pass through the steps might take a day or a week, rather than the months or years of each complete step in the waterfall model. First, one writes automated tests to provide concrete goals for development. Next is coding (by a pair of programmers), which is complete when all the tests pass, and the programmers can't think of any more tests that are needed. Design and architecture emerge out of refactoring, and come after coding. The same people who do the coding do design. (Only the last feature — merging design and code — is common to all the other agile processes.) The incomplete but functional system is deployed or demonstrated for (some subset of) the users (at least one of which is on the development team). At this point, the practitioners start again on writing tests for the next most important part of the system.
- Dynamic system development method.
- Scrum.

3.5 Rapid application development

Rapid application development (RAD) is a software development methodology that uses minimal planning in favor of rapid prototyping (Figure 2-4). The "planning" of software developed using RAD is interleaved with writing the software itself. The lack of extensive preplanning generally allows software to be written much faster and makes it easier to change requirements. RAD involves methods like iterative development and software prototyping. According to Whitten, it is a merger of various structured techniques, especially data-driven information engineering, with prototyping techniques to accelerate software systems development.

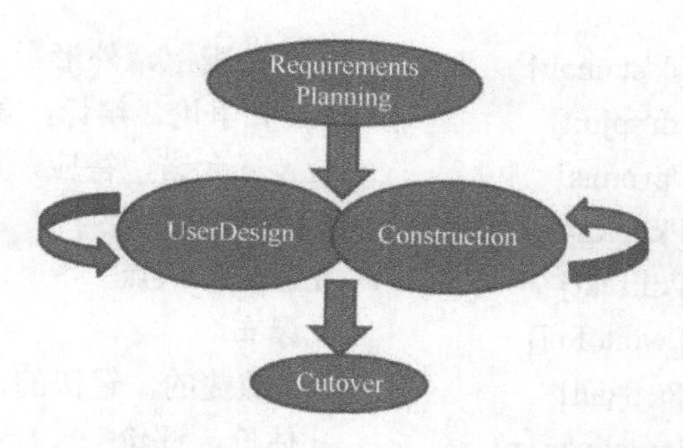

Figure 2-4 Rapid Application Development (RAD) Model

In rapid application development, structured techniques and prototyping are especially used to define users' requirements and to design the final system. The development process starts with the development of preliminary data models and business process models using structured techniques. In the next stage, requirements are verified using prototyping, eventually to refine the data and business process models. These stages are repeated iteratively; further development results in "a combined business requirements and technical design statement to be used for constructing new systems".

3.6 Code and fix

"Code and fix" development is not so much a deliberate strategy as an artifact of naivety and schedule pressure on software developers. Without much of a design in the way, programmers immediately begin producing code. At some point, testing begins (often late in the development cycle), and the unavoidable bugs must then be fixed before the product can be shipped.

New Words

approach	[ə'prəʊtʃ]	*n.*方法，步骤，途径
spiral	['spaɪrəl]	*adj.*螺旋形的 *n.*螺旋
methodology	[ˌmeθə'dɒlədʒɪ]	*n.*方法学，方法论
rating	['reɪtɪŋ]	*n.*评估，评价 *v.*估价；定级(rate 的 ing 形式)
contract	['kɒntrækt]	*n.*合同，契约
predictable	[prɪ'dɪktəbl]	*adj.*可预言的
systematize	['sɪstəmətaɪz]	*v.*系统化
formalize	['fɔ:məlaɪz]	*vt.*使正式，形式化
seemingly	['si:mɪŋlɪ]	*adv.*表面上地
unruly	[ʌn'ru:lɪ]	*adj.*不受拘束的，不守规矩的
practitioner	[præk'tɪʃənə]	*n.*从业者
ambiguous	[æm'bɪgjuəs]	*adj.*不明确的
contradictory	[ˌkɒntrə'dɪktərɪ]	*adj.*反驳的，反对的 *n.*矛盾因素，对立物

externally	[ɪk'stɜ:nəlɪ]	*adv*.外表上，外形上
dispute	[dɪ'spju:t]	*v*.& *n*.争论，辩论，争吵
promise	['prɒmɪs]	*vt*.& *n*.允诺，答应，许诺
clarify	['klærəfaɪ]	*v*.澄清，阐明
defect	['di:fekt]	*n*.过失，缺点
waterfall	['wɔ:təfɔ:l]	*n*.瀑布
agile	['ædʒaɪl]	*adj*.敏捷的，轻快的，灵活的
evaluation	[ɪ,vælju'eɪʃn]	*n*.估价，评价
streamline	['stri:mlaɪn]	*v*.精简，简化，使(系统、机构等)效率更高；尤指使增产节约；便成流线型的
		n.流线，流线型
pro	[prəʊ]	*adv*.正面地
		n.赞成的意见；赞成票，优点
con	[kɒn]	adv.反对地，反面
		n.反对的意见；反对票，缺点
suitable	['su:təbl]	*adj*.适当的，相配的
discourage	[dɪs'kʌrɪdʒ]	*vt*.阻碍
inflexibility	[ɪn,fleksə'bɪlətɪ]	*n*.刚性，非灵活性
neglected	[nɪ'glektɪd]	*adj*.被忽视的
deliberate	[dɪ'lɪbərət]	*adj*.深思熟虑的，故意的，预有准备的
iterative	['ɪtərətɪv]	*adj*.重复的，反复的，迭代的
visualize	['vɪʒuəlaɪz]	*vt*.形象，形象化
quadrant	['kwɒdrənt]	*n*.象限
verification	[,verɪfɪ'keɪʃn]	*n*.确认
condition	[kən'dɪʃn]	*n*.条件，情形，环境
willingness	['wɪlɪŋnəs]	*n*.自愿，乐意
strive	[straɪv]	*v*.努力，奋斗
ignore	[ɪg'nɔ:]	*vt*.不理睬，忽视
incremental	[,ɪŋkrə'mentl]	*adj*.增加的
initialization	[ɪ,nɪʃəlaɪ'zeɪʃn]	*n*.设定初值，初始化
redesign	[,ri:dɪ'zaɪn]	*v*.重新设计
		n.重新设计，新设计
revise	[rɪ'vaɪz]	*vt*.修订，校订，修改
feedback	['fi:dbæk]	*n*.反馈，反应
modularity	[mɒdjʊ'lærɪtɪ]	*n*.模块性
usability	[,ju:zə'bɪlətɪ]	*n*.可用性
efficiency	[ɪ'fɪʃnsɪ]	*n*.效率，功效
achievement	[ə'tʃi:vmənt]	*n*.成就，功绩
advocate	['ædvəkeɪt]	*n*.提倡者，支持者

		*vt.*提倡，支持
fundamentally	[ˌfʌndə'mentəlɪ]	*adv.*基础地，根本地
incomplete	[ˌɪnkəm'pli:t]	*adj.*不完全的，不完善的
concrete	['kɒŋkri:t]	*adj.*具体的，有形的
refactoring	[rɪ'fæktərɪŋ]	*n.*重构
merge	[mɜ:dʒ]	*v.*合并，并入，结合
scrum	[skrʌm]	*n.*橄榄球里的并列争球
accelerate	[ək'seləreɪt]	*v.*加速，促进
artifact	['ɑ:təˌfækt]	*n.*人工制品，手工艺品，加工品
naivety	[naɪ'i:vətɪ]	*n.*天真烂漫，单纯

Phrases

software development process	软件开发过程
systems development life cycle	系统开发生命周期
project management	项目管理
delivery schedule	交付时间表
requirements analysis	需求分析
software testing	软件测试
plan-driven model	计划驱动模型，基于计划的模型
agile model	敏捷模型
waterfall model	瀑布模型
requirements specification	需求规约
spiral model	螺旋模型
risk management	风险管理
rapid prototyping methodology	快速原型法
risk-driven spiral model	风险驱动的螺旋模型
iterative and incremental development	迭代和增量开发
light-weight iterative project	轻量级迭代项目
software design document	软件设计文档
dynamic systems development method	动态系统开发方法
iterative development	迭代开发
data model	数据模型
business process model	业务处理模型

Abbreviations

SDLC (Software Development Life-Cycle)	软件开发生命周期
IEC (International Electrotechnical Commission)	国际电工委员会
SEPG (Software Engineering Process Group)	软件工程过程组
API (Application Program Interface)	应用程序接口

XP (eXtreme Programming)　　极限编程
RAD (Rapid Application Development)　　快速应用开发

Word Building

合成形容词有如下几种形式。

1. 名词+现在分词

peace + loving —— peace-loving 热爱和平的

epoch + making —— epoch-making 划时代的

2. 名词+过去分词

man + made —— man-made 人造的

hand + made —— hand-made 手工制作的

3. 形容词+现在分词

good + looking —— good-looking 好看的

fine + looking —— fine-looking 美观的

deep + going —— deep-going 深入的

4. 形容词+过去分词

ready + made —— ready-made 现成的

5. 形容词+名词

new + type —— new-type 新型的

large + scale —— large-scale 大规模的

6. 形容词+名词+ed

medium + sized —— medium-sized 中型的

noble + minded —— noble-minded 高尚的

7. 形容词+形容词

red + hot —— red-hot 炽热的

light + blue —— light-blue 淡蓝的

dark + green —— dark-green 深绿的

8. 数词+名词

first + class —— first-class 第一流的

three + way —— three-way 三通的,三项的

9. 数词+名词+ed

four + cornered —— four-cornered 有四个角的

one + sided —— one-sided 单面的,片面的

10. 副词+现在分词

hard + working —— hard-working 勤劳的

ever + increasing —— ever-increasing 不断增长的

11. 副词+过去分词

well + known —— well-known 著名的

newly + built —— newly-built 新建的

above + mentioned —— above-mentioned 上述的

12. 介词或副词+名词

under + ground —— underground 地下的

off + hand —— off hand 即刻的

Career Training

求 职 信

得体的求职信是一把金钥匙。求职信一般由 3 部分组成：开头、主体和结尾。

开头部分包括称呼和引言。称呼要恰当，说明应聘缘由和目的；引言的主要作用是尽量引起对方的兴趣，引人注目。

主体部分是求职信的重点，要简明扼要，并具有针对性，突出自己的特点，应努力使自己的描述与所求职位的要求一致，切勿夸大其词或不着边际，外企招聘人员尤其重视这一点。

结尾部分要把想得到工作的迫切心情表达出来，请用人单位尽快答复你并给予面试机会，语气要热情、诚恳、有礼貌。

1）外企求职信一般要用外语写，主要是英语，建议准备好中、英文两份材料。写求职信的过程本身也就反映出了你的外语水平，故应尽量做到语言规范、符合外文习惯，减少语法错误。

2）求职信要有针对性。针对不同企业、不同职位，求职信的内容要有所变化，侧重点要有所不同，使对方觉得你的经历和素质与所聘职位要求一致，因为外企招聘需要的不是最好的员工，而是最适合其所聘职位的人。

3）外企求职信中一定要自信，外企忌讳的是不自信，这与中国传统的“谦虚是美德”略有不同，应充分强调自己的长处和技能，对自己较重要的经历和实践要有较详细的叙述。

4）要本着诚实的原则，不能无中生有，自吹自擂。因为诚实守信是一个人的第一美德。

5）在写外企求职信中，要注意写“怎么干”比“干什么”更重要。例如，你担任过校学生会主席，不要光写头衔，更重要的是写你是如何担任这个职位的，组织了哪些活动，怎么组织这些活动，达到了何种既定目标。因为外企重视的不是你的身份，而是你如何在所任职位上发挥你的才能，他们大多以此来判断你的能力和潜力是否能胜任其所聘职位的工作。

请看如下范例：

Dear Mr. Peterson;

I would like to be considered for the position of Software Engineer in your company. I learned about the position from your listing on the Internet. I will receive my master’s degree in Computer Science from Southeastern University in June.

As indicated by my resume, I have internship experience in both software programming and customer relations. My previous positions involved analyst functions as well as working effectively with customers and clients. My coursework in Computer Science has helped me to focus my career goals on computer engineering. Considering my academic and professional credentials, I am certain that I could make a contribution to your company.

I am very interested in speaking with you about this position and will call you during the first week of March to see if an interview can be arranged. Should you need to reach me in the meantime, please call me at 868-555-6688. Thank you for your consideration.

Sincerely,
Robert Smith

软件水平考试试题解析

【真题再现】

从供选择的答案中选出应填入下列英语文句中____内的正确答案，把编号写在答卷的对应栏内。

Software products may be __A__ into four basic types: application programs, programming language processors, operating systems, and system utilities.

Application programs are programs that __B__ useful tasks such as solving statistical problems, or keeping your company's books.

Programming language processors are programs that __C__ the use of a computer language in a computer system. They are tools for the development of application programs.

Operation systems are programs that __D__ the system resources and enable you to run application programs.

System utilities are special programs that __E__ the usefulness of or add capabilities to a computer.

供选择的答案

A～E：①manage ②perform ③support ④reduce ⑤divided
⑥enhance ⑦implemented ⑧introduce ⑨ranked ⑩run

【答案】A：⑤ B：② C：③ D：① E：⑥

【试题解析】

A：be divided into. 的意思是“被分成……”。本句的意思是“软件产品可分成4种基本类型”。故选⑤。

B: perform 的意思是“做，实行”，与 tasks 搭配时，意思是“执行任务，做工作”；implemented 的意思也是“做，实行”，但它是过去式，时态错误。故选②。

C: 根据后一句“它们是研发应用程序的工具”，可以推出这里应该填含有“支持”意义的词，support 的意思是“支持”。故选③。

D：manage 的意思是“管理”；reduce 的意思是“减少”；enhance 的意思是“增加，提高”；introduce 的意思是“引进，介绍”；ranked 的意思是“排列，归类于”；run 的意思是“运行”。根据句意，应填 manage，意思是“管理资源”。故选①。

E: 根据 add 可以推出，此处应该填一个含有“增加，加强”意义的词。enhance 的意思是“增加、增强、提高”。故选⑥。

【参考译文】

软件产品可分为4种基本类型：应用程序、语言处理程序、操作系统和系统实用程序。

应用程序用来执行应用任务，如解决统计问题，或者帮公司记账。

语言处理程序用来在计算机系统上支持计算机语言的使用。它们是研发应用程序的工具。

操作系统用来管理系统资源、运行应用程序。

系统实用程序用来增强计算机的实用性或提高计算机的性能。

Exercises

[Ex. 1] 根据 Text A 判断正误，正确的标为“T”，错误的标为“F”。

1) Software can be thought of as the invariable part of a computer. (　)

2) Software is often divided into application software and system software. (　　)

3) The utility is a small useful program with unlimited capability. (　)

4) An operating system is initially loaded into the computer by a boot program. (　　)

5) OS handles only the input to and from attached hardware devices, such as hard disks, printers, and dial-up ports. (　　)

6) The application software manages and integrates a computer's capabilities but does not directly perform tasks that benefit the user. (　)

7) Vertical applications are less popular and widespread than horizontal applications. (　　)

8) Applications can also be classified by computing platform, delivery network, or delivery devices. (　)

9) Typically, middleware programs do not provide messaging services. (　　)

10) The firmware of a device may rarely or never be changed during its economic lifetime. (　)

[Ex. 2] 根据 Text B 回答以下问题。

1) What is a software development process?

2) What is the international standard for describing the method of selecting, implementing and monitoring the life cycle for software?

3) Once the general requirements are gathered from the client, what should be determined and clearly stated? What is this often called?

4) When does deployment start? What may it involve?

5) How is the waterfall model commonly taught?

6) What is the key characteristic of a spiral model?

7) What is the basic idea behind iterative and incremental development?

8) What do agile processes fundamentally do?

9) What is rapid application development?

10) What are especially used to define users' requirements and to design the final system in rapid application development?

[Ex. 3] 把下列句子翻译为中文。

1) The company deals in both hardware and software.

2) Your memory is too small to run this application software.

3) Each virtual machine includes the application software and supporting middleware and operating system.

4) The boot program can use any interface to download the application program in the Application Flash Memory.

5) The boot sector is a small program that is the first part of the operating system that the computer loads.

6) All this depends on both the scalability of the database server and the middleware managing the connection pool.

7) Our interest is in the process management and middleware infrastructure.

8) Microsoft is trying to say that Office is relevant and that it can become a development platform that can integrate and operate with other vertical applications to create something new.

9) The following method of firmware update might not be supported in your area or for your hardware.

10) If there's a problem, you have to wait for a "firmware upgrade" -which may or may not address your problem.

[Ex. 4] 软件水平考试真题自测。

从供选择的答案中选出应填入下列英语文句中____内的正确答案，把编号写在答卷的对应栏内。

Software design is a __A__ process. It requires a certain __B__ of flair on the part of the designer. Design cannot be learned from a book. It must be practiced and learnt by experience and study of existing systems. A well __C__ software system is straight-forward to implement and maintain, easily __D__ and reliable. Badly __C__ software system, although they may work, are __E__ to be expensive to maintain, difficult to test and unreliable.

供选择的答案：

A:	①create	②created	③creating	④creative
B:	①amount	②amounted	③mount	④mounted
C:	①design	②designed	③designing	④designs
D:	①understand	②understands	③understanding	④understood
E:	①like	②likely	③unlike	④unlikely

[Ex. 5] 听短文填空。

Exercises 2

File recovery is the process of reconstructing lost or ____1____ files on disk. Files are ____2____ when they are inadvertently deleted, when on-disk information ____3____ their storage is damaged, or when the disk is ____4____. File recovery involves the use of utility ____5____ that attempt to rebuild on-disk information about the storage locations of deleted files. Because deletion makes the file's disk space available but does not ____6____ the data, data that has not yet been overwritten can be ____7____. In the case of damaged files or disks, recovery programs read whatever raw data they can find, and ____8____ the data to a new disk or file in ASCII or numeric (binary or hexadecimal) ____9____. In some instances, however, such reconstructed files contain so much extraneous or mixed information that they are unreadable. The best way to recover a file is to ____10____ it from a backup copy.

Reference Translation

软　件

软件是一个通用术语，是指操作计算机和相关设备所使用的各种程序（术语“硬件”描述的是计算机和相关设备的物理方面）。

软件可以被认为是计算机的可变部分，硬件是不变部分。软件通常分为应用软件（做用户直接关心的工作的程序）和系统软件（包括操作系统和其他所有支持应用软件的程序）。术语“中间件”有时用于描述在应用程序和系统软件之间或两种不同类型的应用软件之间提供中介的程序（例如，把一个远程工作请求从有一种操作系统的计算机的应用程序中发送给另一种操作系统的计算机的应用程序）。

另一种难以分类的软件是实用程序，它是功能有限的小的实用程序。有些实用程序附带有操作系统。就像应用程序一样，实用程序往往是单独安装的，并能够独立于操作系统。

小应用程序（Applet），有时会作为操作系统的“附件”。也可以独立地使用 Java 或其他编程语言创建它们。

软件可以被购买或收购作为共享软件（通常打算试用期后出售）、试用版软件（禁用某些功能的共享软件）、免费软件（免费软件，但带有版权限制）、公共软件（没有任何限制的免费软件）及开源软件（提供软件的源代码，但用户须同意不限制发布改进的版本）。

之前软件通常是打包在 CD-ROM 上的。如今，很多可购买的软件、共享软件和免费软件都通过因特网下载。一个新的趋势是由应用服务供应商的网站来提供软件。

1. 操作系统

操作系统（有时缩写为“OS”）是通过一个引导程序最先加载到计算机并管理计算机中所有其他程序的程序。其他程序被称为应用或应用程序。应用程序通过预设的应用程序接口（API）发送服务请求来使用操作系统。此外，用户可以通过用户接口，例如，一个命令语言或图形用户界面（GUI），直接与操作系统进行交互。

操作系统为应用程序执行如下服务：

- 在能够同时运行多个程序的多任务操作系统中，操作系统决定运行哪些应用程序、以何种顺序运行，以及给每个应用程序多长运行时间，然后运行其他应用程序。
- 管理多个应用程序对内部存储器的共享。
- 处理连接的硬件设备（如硬盘、打印机及拨号端口）的输入和输出。
- 把操作状态和可能发生的任何错误消息发送给每个应用程序或交互式用户（或系统操作员）。
- 卸载所谓的批处理作业（如打印），以便应用程序脱离这项作业。
- 在可以提供并行处理的计算机上，操作系统可以管理如何分解程序，以便在多个处理器中同时运行该程序。

2. 应用软件

应用软件是除了运行计算机本身之外，使计算机执行有用任务的所有计算机软件。这类软件的特定实例被称为软件应用、应用程序、应用或 APP。

该术语与系统软件相对应，系统软件管理和集成了计算机的能力，但不直接执行有益于用户的任务。系统软件服务于应用程序，而应用程序又为用户提供服务。

应用软件包括会计软件、企业软件、图形软件、媒体播放器和办公套件。许多应用程序主要处理的是文档。应用程序可能会与计算机及其系统软件捆绑发布，也可能会单独发布。

应用软件将特定的计算平台或系统软件的强大功能用于某一特定用途。

有些应用程序有可用在多个不同平台的版本，另一些则适用较窄的需求并由需求而得名，例如，用于 Windows 的地理应用程序，用于教育的 Android 应用程序，或用于游戏的 Linux。有时一个新的流行应用程序只能运行在某一平台上，从而增加了该平台的期望。这种应用就是所谓的杀手级应用。

有许多不同的方式对应用软件分类，说明如下：

由于网络的发展和几乎全球化的应用，Web 应用程序之间已经出现了一个重要区别——用 HTML、JavaScript 和其他 Web 原生技术编写的程序，通常需要处于联机状态并运行在 Web 浏览器；而用任何语言编写的传统本地应用程序则用于特定类型的计算机。Web 应用程序是否会取代多种用途的本地应用程序（尤其是在移动设备，如智能手机和平板电脑领域）在计算社区引起争议和辩论；Web 应用程序确实在某些领域中日益流行，但本地应用程序的优势使它们不可能很快消失；两者可以互补，甚至集成。

也可以认为，应用软件要么是横向的，要么是垂直的。横向应用软件更普及和广泛，例如，文字处理器或数据库。垂直应用软件是细分产品，专为特定类型的行业、业务或组织内的特定部门而设计。软件集成套件将尽力适应每一个可能的具体方面，例如，制造、银行系统、会计或客户服务。

应用软件有许多类型：

- 应用软件套件包括多个捆绑在一起的应用软件。它们通常具有相关的功能、特征和用户界面，并且可交互，例如，打开对方的文件。业务应用软件常常被做成套件，例如，微软 Office、LibreOffice 和 iWork，其中捆绑了文字处理器、电子表格等；但套件也有其他用途，例如，用于图形或音乐。
- 企业级软件满足整个组织中几乎所有部门的流程和数据流的需求，往往在一个大的分布式环境中，如财务系统、客户关系管理系统和供应链管理系统。部门软件是企业软件的一个子类，专注于大型组织内较小的组织和/或组，如差旅费管理和 IT 服务支持。
- 企业基础架构软件提供支持企业级软件系统所需的通用功能，如数据库、电子邮件服务器及用于管理网络和安全的系统。
- 信息工作者软件允许用户创建和管理（往往是一个部门内的个别项目）信息，与企业管理相对应，包括时间管理、资源管理、文档工具、分析和协作等软件。在多个信息工作者任务中还有文字处理器、电子表格、电子邮件和博客客户端、个人信息系统及个别媒体的编辑器等软件。
- 内容访问软件主要用于访问无须编辑的内容，但也可能包括允许内容编辑的软件。这样的软件满足了个人和团体对数字娱乐和数字出版的需求，如媒体播放器、网络浏览器和帮助浏览器。
- 教育软件与内容访问软件相关，但内容和/或功能适用于教育工作者或学生。例如，它可以提供评估（测试）、通过材料跟踪进度。
- 仿真软件模拟物理或抽象的系统，可用于研究、培训，也可用于娱乐。
- 媒体开发软件生成平面媒体和电子媒体以供他人使用，最常见于商业或教育环境。这类软件包括图形艺术软件、排版软件、多媒体开发软件、HTML 编辑器、数字动画编辑器、数字音频和视频合成及其他众多应用软件。
- 产品工程软件用于开发硬件和软件产品。这类软件包括计算机辅助设计（CAD）、计算机辅助工程（CAE）、计算机语言编辑和编译工具、集成开发环境和应用程序接口。

应用软件也可以按照以下维度来分类：计算平台，如一个特定的操作系统；分发网络，如云计算和 Web 2.0 应用软件；分发设备，如用于移动设备的移动 App。

执行简单计算、测量、绘制和文字处理任务时，操作系统本身可以被当作应用软件。这类软件不包括操作系统中捆绑的应用软件，如软件计算器或文本编辑器。

3. 中间件

在计算机行业，中间件是一个通用术语，用于表示把两个独立的、往往已经存在的程序“黏合在一起”或提供中介的任一程序。中间件的一个常见应用是允许为访问特定数据库而编写的程序访问其他数据库。

通常情况下，中间件程序提供消息服务，以便不同的应用程序间可以进行通信。不同的应用程序通常通过中间件组合成一个系统，即“企业应用集成”（EAI）。

4. 固件

在电子系统和计算科学中，固件是永久内存、程序代码和数据（存储在固件中的数据）的组合。包含固件设备的典型例子有嵌入式系统（如交通信号灯、家用电器及数字手表）、计算机、计算机外部设备、手机、数码相机。包含在这些设备中的固件提供该设备的控制程序。固件保存在非易失性存储器装置（例如ROM、EPROM或闪存）中。设备中的固件在其经济寿命期间很少更改或者从不更改。一些固件存储设备永久安装，出厂后不能更改。更新固件的常见原因包括修复错误或给设备添加功能。这可能需要在物理上改变 ROM 的集成电路，或用特殊的程序对闪存重新编程。如个人计算机的 ROM 中的 BIOS，这类固件可能仅包含设备的初级基本功能，并且只对高级软件提供服务；而如嵌入式系统的程序，这类固件也许是在系统上运行的唯一程序，并提供其全部功能。

Unit 3

Text A

Program Design

1. Program

In computing, a program is a specific set of ordered operations for a computer to perform. In the modern computer that John Von Neumann outlined in 1945, the program contains a one-at-a-time sequence of instructions that the computer follows.[1] Typically, the program is put into a storage area accessible to the computer. The computer gets one instruction and performs it and then gets the next instruction. The storage area or memory can also contain the data that the instruction operates on. (Note that a program is also a special kind of "data" that tells how to operate on "application or user data.")

Programs can be characterized as interactive or batch in terms of what drives them and how continuously they run. An interactive program receives data from an interactive user (or possibly from another program that simulates an interactive user). A batch program runs and does its work, and then stops. Batch programs can be started by interactive users who request their interactive program to run the batch program. A command interpreter or a Web browser is an example of an interactive program. A program that computes and prints out a company payroll is an example of a batch program. Print jobs are also batch programs.

When you create a program, you write it using some kind of computer language. Your language statements are the source program. You then "compile" the source program (with a special program called a language compiler) and the result is called an object program (not to be confused with object-oriented programming). There are several synonyms for object program, including object module and compiled program. The object program contains the string of 0s and 1s called machine language that the logic processor works with.

The machine language of the computer is constructed by the language compiler with an understanding of the computer's logic architecture, including the set of possible computer instructions and the length (number of bits) in an instruction.

2. Structured Programming (Modular Programming)

Structured programming (sometimes known as modular programming) is a subset of procedural programming that enforces a logical structure on the program being written to make it more efficient and easier to understand and modify. Certain languages such as Ada, and Pascal are designed with features that encourage or enforce a logical program structure.

Structured programming frequently employs a top-down design model, in which developers map out the overall program structure into separate subsections.[2] A defined function or set of similar functions is coded in a separate module or submodule, which means that code can be loaded into memory more efficiently and that modules can be reused in other programs.[3] After a module has been tested individually, it is then integrated with other modules into the overall program structure.

Program flow follows a simple hierarchical model that employs looping constructs such as "for" "repeat" and "while". Use of the "Go To" statement is discouraged.

Structured programming was first suggested by Corrado Bohm and Guiseppe Jacopini. The two mathematicians demonstrated that any computer program can be written with just three structures: sequences, decisions, and loops. Edsger Dijkstra's subsequent article, Go To Statement Considered Harmful, was instrumental in the trend towards structured programming. The most common methodology employed was developed by Dijkstra. In this model (which is often considered to be synonymous with structured programming, although other models exist) the developer separates programs into subsections that each have only one point of access and one point of exit.

Almost any language can use structured programming techniques to avoid common pitfalls of unstructured languages. Unstructured programming must rely upon the discipline of the developer to avoid structural problems, and as a consequence may result in poorly organized programs. Most modern procedural languages include features that encourage structured programming. Object-oriented programming (OOP) can be thought of as a type of structured programming. It uses structured programming techniques for program flow, and adds more structure for data to the model.

3. Compiler

A compiler is a special program that processes statements written in a particular programming language and turns them into machine language or "code" that a computer's processor uses. Typically, a programmer writes language statements in a language such as Pascal or C one line at a time using an editor. The file that is created contains what are called the source statements. The programmer then runs the appropriate language compiler and specifies the name of the file that contains the source statements.

When executing (running), the compiler first parses (or analyzes) all of the language statements syntactically one after another and then, in one or more successive stages or "passes", builds the output code, making sure that statements that refer to other statements are referred to correctly in the final code. Traditionally, the output of the compilation has been called object code or sometimes an object module. (Note that the term "object" here is not related to object-oriented programming.) The object code is machine code that the processor can process or "execute" one instruction at a time.

More recently, the Java programming language, a language used in object-oriented programming, has introduced the possibility of compiling output (called bytecode) that can run on any computer system platform for which a Java virtual machine or bytecode interpreter is provided to convert the bytecode into instructions that can be executed by the actual hardware processor.[4]

Using this virtual machine, the bytecode can optionally be recompiled at the execution platform by a just-in-time compiler .

Traditionally in some operating systems, an additional step is required after compilation that of resolving the relative location of instructions and data when more than one object module is to be run at the same time and they cross-referred to each other's instruction sequences or data. This process is sometimes called linkage editing and the output known as a load module.

A compiler works with what are sometimes called 3GL and higher-level languages. An assembler works on programs written using a processor's assembler language.

4. Interpreter

An interpreter is a program that executes instructions written in a high-level language. There are two ways to run programs written in a high-level language. The most common is to compile the program; the other method is to pass the program through an interpreter.

An interpreter translates high-level instructions into an intermediate form, which it then executes. In contrast, a compiler translates high-level instructions directly into machine language. Compiled programs generally run faster than interpreted programs. The advantage of an interpreter, however, is that it does not need to go through the compilation stage during which machine instructions are generated. This process can be time consuming if the program is long. The interpreter, on the other hand, can immediately execute high-level programs. For this reason, interpreters are sometimes used during the development of a program, when a programmer wants to add small sections at a time and test them quickly. In addition, interpreters are often used in education because they allow students to program interactively.

Both interpreters and compilers are available for most high-level languages. However, BASIC and LISP are especially designed to be executed by an interpreter. In addition, page description languages, such as PostScript, use an interpreter. Every PostScript printer, for example, has a built-in interpreter that executes PostScript instructions.

5. High Level Language

High level languages are similar to the human language. Unlike low level languages, high level languages are programmers friendly, easy to code, debug and maintain.

High level language provides higher level of abstraction from machine language. They do not interact directly with the hardware. Rather, they focus more on the complex arithmetic operations, optimal program efficiency and easiness in coding.

High level programs require compilers/interpreters to translate source code into machine language. We can compile the source code written in high level language to multiple machine languages. Thus, they are machine independent language.

Today almost all programs are developed using a high level programming language. A variety of applications are developed using high level language. High level languages are used to develop desktop applications, websites, system software, utility software and many more.

High level languages are grouped in two categories based on execution model: compiled languages and interpreted languages.

6. Scripting Language

A scripting language is a high-level programming language that is interpreted by another program at runtime rather than compiled by the computer's processor as other programming languages (such as C and C++) are.[5] Scripting languages, which can be embedded within HTML, are commonly used to add functionality to a Web page, such as different menu styles or graphic displays or to serve dynamic advertisements. These types of languages are client-side scripting languages and they affect the data that the end user sees in a browser window. Other scripting languages are server-side scripting languages that manipulate the data, usually in a database, on the server.

Scripting languages came about largely because of the development of the Internet as a communications tool. JavaScript, ASP, JSP, PHP, Perl, Tcl and Python are examples of scripting languages.

New Words

storage	['stɔːrɪdʒ]	*n*.存储
accessible	[ək'sesəbl]	*adj*.可访问的
characterize	['kærəktəraɪz]	*vt*.表现……的特色
interpreter	[ɪn'tɜːprɪtə]	*n*.解释程序，解释器
statement	['steɪtmənt]	*n*.语句
module	['mɒdjuːl]	*n*.模块
construct	[kən'strʌkt]	*vt*.建造，构造，创立
procedural	[prə'siːdʒərəl]	*adj*.过程化的，程序上的
modify	['mɒdɪfaɪ]	*vt*.更改，修改
encourage	[ɪn'kʌrɪdʒ]	*vt*.鼓励
subsection	['sʌbsekʃn]	*n*.分部，分段，小部分，小单位
submodule	[sʌb'mɒdjuːl]	*n*.子模块
demonstrate	['demənstreɪt]	*vt*.证明，论证
decision	[dɪ'sɪʒn]	*n*.决策
subsequent	['sʌbsɪkwənt]	*adj*.后来的
instrumental	[ˌɪnstrə'mentl]	*adj*.有帮助的，起作用的
developer	[dɪ'veləpə]	*n*.开发者
unstructured	[ʌn'strʌktʃəd]	*adj*.非结构化的，无结构的
consequence	['kɒnsɪkwəns]	*n*.结果，推论，因果关系
poorly	['pʊəlɪ]	*adj*.恶劣的
editor	['edɪtə]	*n*.编辑器
execute	['eksɪkjuːt]	*vt*.执行，实行
parse	[pɑːz]	*vt*.解析
analyze	['ænəlaɪz]	*vt*.分析，分解
syntactically	[sɪn'tæktɪklɪ]	*adv*.依照句法地，在语句构成上
successive	[sək'sesɪv]	*adj*.继承的，连续的

compilation	[ˌkɒmpɪ'leɪʃn]	*n.*编译
bytecode	['baɪtkəʊd]	*n.*字节码
convert	[kən'vɜ:t]	*vt.*使转变，转换……
actual	['æktʃuəl]	*adj.*实际的，真实的，现行的
recompile	[rɪkəm'pɑɪl]	*vt.*重新编译
linkage	['lɪŋkɪdʒ]	*n.*连接，联结
method	['meθəd]	*n.*方法
translate	[træns'leɪt]	*vt.*翻译，解释，转化
built-in	[bɪlt-ɪn]	*adj.*内置的，固定的，嵌入的 *n.*内置
script	[skrɪpt]	*n.*脚本
functionality	[ˌfʌŋkʃə'næləti]	*n.*功能性
dynamic	[daɪ'næmɪk]	*adj.*动态的
manipulate	[mə'nɪpjuleɪt]	*vt.*操作，使用

Phrases

one at a time	一次一个
storage area	存储区
batch program	批处理程序
print job	打印作业，打印任务
source program	源程序
object program	目标程序
be confused with …	与……混淆
object-oriented programming	面向对象程序设计
machine language	机器语言
structured programming	结构化程序设计
modular programming	模块化程序设计
top-down design model	自顶向下的设计模型
map out	描绘出
separates … into …	把……分为……
program flow	程序流程
turns … into …	把……转变为……
object code	目标代码
object module	目标模块
Java virtual machine	Java 虚拟机
convert … into …	把……转换为……
just-in-time compiler	即时编译器
relative location	相对位置
load module	载入模块

higher-level language	更高级的语言
time consuming	耗费时间的
page description language	页面描述语言
scripting language	脚本语言
Web page	网页
client-side scripting language	客户端脚本语言
server-side scripting language	服务器端脚本语言

Abbreviations

OOP (Object-Oriented Programming)	面向对象编程
3GL (Third-Generation Language)	第三代编程语言
BASIC (Beginners All-purpose Symbolic Instruction Code)	初学者通用指令码
LISP (LISt Processor)	列表处理语言
COBOL (COmmon Business-Oriented Language)	面向商业的通用语言
FORTRAN (FORmula TRANslation)	公式翻译语言
ASP (Active Server Pages)	活动服务器页面
JSP (Java Server Pages)	Java 服务器页面

Notes

[1] In the modern computer that John Von Neumann outlined in 1945, the program contains a one-at-a-time sequence of instructions that the computer follows.

本句中，that John Von Neumann outlined in 1945 是一个定语从句，修饰和限定 the modern computer。that the computer follows 也是一个定语从句，修饰和限定 instructions。one-at-a-time 的意思是“一次一个”。

[2] Structured programming frequently employs a top-down design model, in which developers map out the overall program structure into separate subsections.

本句中，in which developers map out the overall program structure into separate subsections 是一个介词前置的非限定性定语从句，对 a top-down design model 进行补充说明。

[3] A defined function or set of similar functions is coded in a separate module or submodule, which means that code can be loaded into memory more efficiently and that modules can be reused in other programs.

本句中，which means that code can be loaded into memory more efficiently and that modules can be reused in other programs 是一个非限定性定语从句，对它前面的整个句子进行补充说明。

[4] More recently, the Java programming language, a language used in object-oriented programming, has introduced the possibility of compiling output (called bytecode) that can run on any computer system platform for which a Java virtual machine or bytecode interpreter is provided to convert the bytecode into instructions that can be executed by the actual hardware processor.

本句中，a language used in object-oriented programming是 the Java programming language

的同位语，对其进行解释说明。used in object-oriented programming是一个过去分词短语，作a language 的定语。that can run on any computer system platform for which a Java virtual machine or bytecode interpreter is provided to convert the bytecode into instructions that can be executed by the actual hardware processor 是一个定语从句，修饰和限定 compiling output (called bytecode)；在该从句中，for which a Java virtual machine or bytecode interpreter is provided 是一个介词前置的定语从句，修饰和限定 any computer system platform；that can be executed by the actual hardware processor 是一个定语从句，修饰和限定 instructions。

[5] A scripting language is a high-level programming language that is interpreted by another program at runtime rather than compiled by the computer's processor as other programming languages (such as C and C++) are.

本句中，that is interpreted by another program at runtime rather than compiled by the computer's processor as other programming languages (such as C and C++) are 是一个定语从句，修饰和限定 a high-level programming language。rather than 的意思是"而不是"。

Text B

Data, Data Structure and Database

1. Data

(1) Distinct pieces of information, usually formatted in a special way. All software is divided into two general categories: data and programs. Programs are collections of instructions for manipulating data.

Data can exist in a variety of forms-as numbers or text on pieces of paper, as bits and bytes stored in electronic memory, or as facts stored in a person's mind.

Strictly speaking, data is the plural of datum, a single piece of information. In practice, however, people use data as both the singular and plural form of the word.

(2) The term data is often used to distinguish binary machine-readable information from textual human-readable information. For example, some applications make a distinction between data files (files that contain binary data) and text files (files that contain ASCII data).

(3) In database management systems, data files are the files that store the database information, whereas other files, such as index files and data dictionaries, store administrative information, known as metadata.

2. Data Structure

In computer science, a data structure is a particular way of storing and organizing data in a computer so that it can be used efficiently.

Different kinds of data structures are suited to different kinds of applications, and some are highly specialized to specific tasks. For example, B-trees are particularly well-suited for implementation of databases, while compiler implementations usually use hash tables to look up identifiers.

Data structures provide a means to manage large amounts of data efficiently, such as large databases and internet indexing services. Usually, efficient data structures are a key to designing efficient algorithms. Some formal design methods and programming languages emphasize data structures, rather than algorithms, as the key organizing factor in software design. Storing and retrieving can be carried out on data stored in both main memory and in secondary memory.

2.1 Basic types

- An array stores a number of elements in a specific order. They are accessed using an integer to specify which element is required (although the elements may be of almost any type). Arrays may be fixed-length or expandable.
- Records are among the simplest data structures. A record is a value that contains other values, typically in fixed number and sequence and typically indexed by names. The elements of records are usually called fields or members.
- A hash table (also called a dictionary or map) is a more flexible variation on a record, in which name-value pairs can be added and deleted freely.
- A union type specifies which of a number of permitted primitive types may be stored in its instances, e.g. float or long integer. Contrast with a record, which could be defined to contain a float and an integer; whereas, in a union, there is only one value at a time.
- A tagged union (also called a variant, variant record, discriminated union, or disjoint union) contains an additional field indicating its current type, for enhanced type safety.
- A set is an abstract data structure that can store specific values, without any particular order, and with no repeated values. Values themselves are not retrieved from sets, rather one tests a value for membership to obtain a boolean "in" or "not in".
- Graphs and trees are linked abstract data structures composed of nodes. Each node contains a value and also one or more pointers to other nodes. Graphs can be used to represent networks, while trees are generally used for sorting and searching, having their nodes arranged in some relative order based on their values.
- An object contains data fields, like a record, and also contains program code fragments for accessing or modifying those fields. Data structures not containing code, like those above, are called plain old data structures.

Many others are possible, but they tend to be further variations and compounds of the above.

2.2 Basic principles

Data structures are generally based on the ability of a computer to fetch and store data at any place in its memory, specified by an address—a bit string that can be itself stored in memory and manipulated by the program. Thus the record and array data structures are based on computing the addresses of data items with arithmetic operations; while the linked data structures are based on storing addresses of data items within the structure itself. Many data structures use both principles, sometimes combined in non-trivial ways (as in XOR linking).

The implementation of a data structure usually requires writing a set of procedures that create and manipulate instances of that structure. The efficiency of a data structure cannot be analyzed separately from those operations. This observation motivates the theoretical concept of an abstract data type, a data structure that is defined indirectly by the operations that may be performed on it, and the mathematical properties of those operations (including their space and time cost).

3. Database

Often abbreviated DB. A collection of information organized in such a way that a computer program can quickly select desired pieces of data. You can think of a database as an electronic filing system.

Traditional databases are organized by fields, records, and files. A field is a single piece of information; a record is one complete set of fields; and a file is a collection of records. For example, a telephone book is analogous to a file. It contains a list of records, each of which consists of three fields: name, address, and telephone number.

An alternative concept in database design is known as hypertext. In a hypertext database, any object, whether it be a piece of text, a picture, or a film, can be linked to any other object. Hypertext databases are particularly useful for organizing large amounts of disparate information, but they are not designed for numerical analysis.

To access information from a database, you need a database management system (DBMS). This is a collection of programs that enables you to enter, organize, and select data in a database.

4. Database Management System

A collection of programs that enables you to store, modify, and extract information from a database. There are many different types of DBMSs, ranging from small systems that run on personal computers to huge systems that run on mainframes. The following are examples of database applications:

- Computerized library systems.
- Automated teller machines.
- Flight reservation systems.
- Computerized parts inventory systems.

From a technical standpoint, DBMSs can differ widely. The terms relational, network, flat, and hierarchical all refer to the way a DBMS organizes information internally. The internal organization can affect how quickly and flexibly you can extract information.

Requests for information from a database are made in the form of a query, which is a stylized question. For example, the query

SELECT ALL WHERE NAME = "SMITH" AND AGE > 35

requests all records in which the NAME field is SMITH and the AGE field is greater than 35. The set of rules for constructing queries is known as a query language. Different DBMSs support different query languages, although there is a semi-standardized query language called SQL (structured query language). Sophisticated languages for managing database systems are called fourth-generation languages, or 4GLs for short.

The information from a database can be presented in a variety of formats. Most DBMSs include a report writer program that enables you to output data in the form of a report. Many DBMSs also include a graphics component that enables you to output information in the form of graphs and charts.

5. RDBMS

Short for relational database management system and pronounced as separate letters, a type of database management system (DBMS) that stores data in the form of related tables. Relational databases are powerful because they require few assumptions about how data is related or how it will be extracted from the database. As a result, the same database can be viewed in many different ways.

An important feature of relational systems is that a single database can be spread across several tables. This differs from flat-file databases, in which each database is self-contained in a single table.

Almost all full-scale database systems are RDBMS's. Small database systems, however, use other designs that provide less flexibility in posing queries.

New Words

distinct	[dɪ'stɪŋkt]	*adj.*清楚的，明显的
machine-readable	[mə'ʃi:n-'ri:dəbəl]	*adj.*计算机可读的，可用计算机处理的
human-readable	['hju:mən-'ri:dəbl]	*adj.*人可读的
index	['ɪndeks]	*n.*索引
		*vt.*编入索引中
		*vi.*做索引
metadata	['metədeɪtə]	*n.*元数据
particularly	[pə'tɪkjuləlɪ]	*adv.*独特地，显著地
identifier	[aɪ'dentɪfaɪə]	*n.*标识符
emphasize	['emfəsaɪz]	*vt.*强调，着重
factor	['fæktə]	*n.*因素，要素
expandable	[ɪk'spændəbl]	*adj.*可扩展的，可扩大的
fixed	[fɪkst]	*adj.*固定的，确定的，准备好的
map	[mæp]	*n.*映射
flexible	['fleksəbl]	*adj.*柔韧性，灵活的
union	['ju:nɪən]	*n.*联合，合并，结合
primitive	['prɪmətɪv]	*adj.*原始的，简单的
float	[fləʊt]	*n.*浮点
define	[dɪ'faɪn]	*vt.*定义，详细说明
discriminate	[dɪ'skrɪmɪneɪt]	*v.*区别
disjoint	[dɪs'dʒɔɪnt]	*v.*（使）脱节，（使）解体
abstract	['æbstrækt]	*adj.*抽象的，理论的

		n.摘要，概要，抽象
compose	[kəm'pəʊz]	*v*.组成
pointer	['pɔɪntə]	*n*.指针
sorting	['sɔ:tɪŋ]	*n*.排序
arrange	[ə'reɪndʒ]	*v*.安排，排列
fragment	['frægmənt]	*n*.碎片，断片，片段
motivate	['məʊtɪveɪt]	*v*.激发，促进
piece	[pi:s]	*n*.块，件，片
alternative	[ɔ:l'tɜ:nətɪv]	*n*.可供选择的办法（事物）
		adj.选择性的
hypertext	['haɪpətekst]	*n*.超文本
disparate	['dɪspərət]	*adj*.全异的
computerize	[kəm'pju:təraɪz]	*vt*.用计算机处理，使计算机化
standpoint	['stændpɔɪnt]	*n*.立场，观点
hierarchical	[ˌhaɪə'rɑ:kɪkl]	*adj*.分等级的，分层的
query	['kwɪərɪ]	*n*.&*v*.查询
semi-standardized	['semɪ-'stændədaɪzd]	*adj*.半标准的
chart	[tʃɑ:t]	*n*.图表
		vt.制图

Phrases

divide …into…	把……分为……
be suited to	适合
data file	数据文件
index file	索引文件
data dictionary	数据字典
data structure	数据结构
hash table	哈希（散列）表
look up	查找
secondary memory	辅助存储器
long integer	长整型
tagged union	标签联合
discriminated union	可区分联合
disjoint union	分割联合
data field	数据域，数据字段
plain old data structure	普通传统数据结构
bit string	位串
numerical analysis	数字分析
flight reservation system	机票预定系统

computerized parts inventory system 计算机零部件库房管理系统
query language 查询语言
flat-file database 平面文件数据库

Abbreviations

B-tree (Binary Tree) 二叉树
XOR (Exclusive OR) 异或（逻辑运算）
DB (DataBase) 数据库
DBMS (DataBase Management System) 数据库管理系统
SQL (Structured Query Language) 结构化查询语言
4GL (Fourth-Generation Language) 第四代语言
RDBMS (Relational DataBase Management System) 关系数据库管理系统

Word Building

合成动词有以下几种形式：
1. 名词+动词
work + harden —— work-harden 加工硬化
heat + treat —— heat-treat 热处理
trial + produce —— trial-produce 试制
2. 形容词+动词
safe + guard —— safeguard 保卫
white + wash —— whitewash 刷白
3. 副词或介词+动词
over + heat —— overheat 过热
over + write —— overwrite 覆盖
up + set —— upset 推翻
under + line —— underline 在……下面画线

Career Training

个 人 简 历

在求职（job-hunting）过程中，求职者通常有必要写一份个人简历，来介绍个人的基本情况。一份好的简历对求职成功起着至关重要的作用。简历通常包括名字、住址、电话号码、个人经历、教育程度（education）、个人基本情况（personal data）、工作目的（job objective）、个人总结（summary）、公开出版物（publication）、外语水平（foreign language skills）、专业水平（professional membership）、爱好（hobbies）、介绍信（references）等。其中，名字、住址、电话号码、个人经历、教育程度、个人基本情况是简历中必不可少的部分。

请看以下范例：

RESUME

Mingjun Liu 506 Chang An St. #286 Shanghai, China. Post Code 200000 Home Phone: (021)-88665518 Email:liumingjun@126.com	
Objective:	
A position as a sales manager for computer company.	
Education:	
2013—2017	Zhejiang University, Hangzhou, China. Bachelor of Science in Computer Science.
Experience:	
2019—present	as a vice sales manager in a computer company.
2017—2019	as an assistant to an administrator in a computer company.
Foreign Language Proficiency:	
CET6；BEC4	
Personal Data:	
Date of Birth: March 16, 1989 Place of Birth: Hangzhou, Zhejiang Province, China Nationality: Chinese Sex: Male Health: Excellent Hobbies: Football and tennis	
References	
Available upon request	

软件水平考试试题解析

【真题再现】

从供选择的答案中选出应填入下列英语文句中____内的正确答案，把编号写在答卷的对应栏内。

The C programming language has __A__ one of the most __B__ programming languages, and it has been implemented on most personal computers and multiuser systems, especially those designed for research and development。It evolved from the version described in Kernighan and Ritche's work (called "K&R C" after the authors) into __C__ variants, including the standard ANSI C, which __D__ many type-checking features and includes a standard library. Of the two main __E__, K&R C is probably the most commonly used on multiuser computers, with ANSI C close behind：In the personal computing world, ANSI C is far more common.

供选择的答案

A～E：①much ②variants ③complex ④incorporates ⑤several ⑥become ⑦popular ⑧editions ⑨come ⑩users

【答案】 A：⑥ B：⑦ C：⑤ D：④ E：②

【试题解析】

A：根据句子结构和句意，此处应该填一个动词的现在分词且意思是“成为、变成”，

become 的意思是“成为、变成”。故选⑥。

B: 此处应填一个形容词，complex 的意思是“复杂的”；several 的意思是“数个的，几个”；popular 的意思是“流行的，受欢迎的”。根据句意，此处应选“流行的，受欢迎的”。故选⑦。

C: 此处应填一个形容词，根据句意，此处应选“数个的，几个”。故选⑤。

D: 根据句子结构，此处应该填一个动词，再根据句意，应填 incorporates，意思是“并入，引入”。故选④。

E: 根据句子结构，此处应该填一个名词，再根据句意，应填 variants，意思是“变体”。故选②。

【参考译文】

C 语言已成为最流行的编程语言之一，多用于个人计算机和多用户系统，尤其是那些为研究和开发而设计的系统。它起源于 Kernighan 和 Ritche 著作中描述的版本（被作者称为“K&R C”），现在已发展为多个版本，包括标准 ANSI C。标准 ANSI C 引入了类型检查特性，包含标准库。两种版本的主要不同之处在于，K&R C 可能在多用户计算机系统中用得最多，而标准 ANSI C 紧随其后，ANSI C 在个人计算机上更为普及。

Exercises

[Ex. 1] 根据 Text A 判断正误，正确的标为“T”，错误的标为“F”。

1) Programs can be characterized as interactive or batch and a command interpreter or a Web browser is an example of an interactive program. (　　)

2) Structured programming frequently employs a bottom-up design model, in which developers map out separate subsections into the overall program structure. (　　)

3) Almost all languages can use structured programming techniques to avoid common pitfalls of unstructured languages. (　　)

4) A compiler is a special program that processes statements written in a particular programming language and turns them into machine language or “code” that a computer’s processor uses. (　　)

5) A compiler works with 3GL and higher-level languages and on programs written using a processor’s assembler language. (　　)

6) Interpreted programs generally don’t run as fast as compiled programs. (　　)

7) An interpreter translates high-level instructions into machine language while a compiler translates high-level instructions directly into an intermediate form. (　　)

8) High level programs require compilers to translate source cade into machine languages. (　　)

9) A scripting language is a low-level programming language that is interpreted by another program at runtime. (　　)

10) Scripting languages can be embedded within HTML. (　　)

[Ex. 2] 根据 Text B 回答以下问题。

1) How many general categories is software divided into?

2) What is a data structure in computer science?

3) What is a record?

4) What is a set?

5) What are graphs and trees? What are they used to do?

6) What does the implementation of a data structure usually require?

7) What are a field, a record and a file respectively?

8) What is a database management system?

9) What are the examples of database applications mentioned in the passage?

10) Why are relational databases powerful?

[Ex. 3] 把下列句子翻译为中文。

1) A virus is a self-replicating computer program that spreads itself into other programs, often without the knowledge or consent of the computer owner.

2) Google Maps is an interactive program for most smart phones that offers tons of features so you will never have to be lost again.

3) Batch and utility programs typically deal with large amounts of data, often processing the data in a sequential manner.

4) Structured Programming utilizes sequential, selective, and repetitive structures to design programs.

5) C language is a general-purpose programming language that provides code efficiency, elements of structured programming, and a rich set of operators.

6) By compiling before the program executes, the AOT compiler must be conservative about classes, fields, and methods referenced by the code it compiles.

7) If you step through the code you will see that the compiler is very flexible when it creates these methods.

8) Interpreted programs are much slower than compiled or assembled ones because a statement must be translated with each execution.

9) In many scripting languages, you don't have to worry about how memory is managed, but that doesn't make memory management any less important.

10) The drawback of scripting languages has always been performance.However, scripting languages have distinct advantages, too.

[Ex. 4] 软件水平考试真题自测。

从供选择的答案中选出应填入下列英语文句中____内的正确答案，把编号写在答卷的对应栏内。

PCs originated as stand-alone __A__ , however, in recent years many have been __B__ to Local Area Networks (LANs). In a LAN, the data and usually the user application reside on the File Server, a PC running a special Network Operating System (NOS) such as Novell's NetWare or Microsoft's LAN Manager. The File Server manages the LAN users' shared access to data on its hard __C__ and frequently provides access to other shared resources, such as printers. While a LAN enables users of PC-based databases to share __D__ data files, it doesn't significantly change how the DBMS works; all the actual data __E__ is still performed on the PC running the database application.

供选择的答案:

A～E：①calculating ②common ③connected ④disks ⑤displayed
⑥systems ⑦printers ⑧processing ⑨some ⑩workstations

[Ex. 5] 听短文填空。

Exercises 3

Object-oriented programming (OOP) is a programming language model in which programs are organized around data, or objects, rather than functions and logic. An object can be defined as a data ____1____ that has unique attributes and behavior. Examples of an object can range from ____2____ entities, such as a human being that is described by properties like ____3____ and address, down to small computer ____4____, such as widgets. This opposes the historical approach to programming where emphasis was placed on how the logic was written rather than how to ____5____ the data within the logic.

Simply put, OOP focuses on the ____6____ that developers want to manipulate rather than the logic ____7____ to manipulate them. This approach to programming is well-suited for programs that are large, ____8____ and actively updated or maintained. Due to the organization of an object-oriented program, this ____9____ is also conducive to collaborative development where projects can be ___10___ into groups. Additional benefits of OOP include code reusability, scalability and efficiency.

Reference Translation

程 序 设 计

1. 程序

在计算科学中，程序是一组特定的由计算机来执行的命令。在 1945 年由约翰·冯·诺伊曼描绘的现代计算机中，程序包含计算机可执行的指令序列，一次执行一个指令。通常程序被放入一个计算机可读取的存储区域。计算机得到一个指令，执行它，然后得到下一个指令。存储区或存储器中也可以包含指令所用的数据（请注意，程序也是一种特殊的数据，它指明如何运行应用程序或用户数据）。

就如何驱动以及怎样连续地运行程序而言，程序可以具有交互或批处理的特点。一个交互式程序从交互式用户处接收数据（或者可能从模拟交互式用户的另一个程序处接收数据）。批处理程序运行并完成其工作，然后停止。那些请求自己的交互式程序运行批处理程序的交互用户可以启动批处理程序。命令解释器或 Web 浏览器就是交互式程序的例子，计算并打印出某个公司工资的程序是批处理程序的一个例子，打印作业也是批处理程序。

当创建一个程序时，你可以使用某种计算机语言来编写。语句段就是源程序。然后，编译源程序（用一个叫作语言编译器的特殊程序），其结果被称为目标程序（不要与面向对象编程混淆）。目标程序有几个同义词，包括目标模块和编译后的程序。目标程序中包含被称为机器语言的供逻辑处理器使用的一系列 0 和 1。

计算机的机器语言由能够理解计算机逻辑体系结构的语言编译器构建，其中包括一组可能的计算机指令和一个指令的长度（位数）。

2. 结构化程序设计（模块化程序设计）

结构化程序设计（有时也被称为模块化程序设计）是过程式程序设计的一个子集，增强了程序的逻辑结构，使其更有效、更容易理解和修改。某些语言（如Ada和Pascal）的设计具有鼓励或强制使用逻辑程序结构的特点。

结构化程序设计中经常采用自顶向下的设计模型，其中开发人员把整个程序结构分成独立的小节。把定义好的功能或一组类似的功能编码为一个单独的模块或子模块，这意味着代码可以更有效地加载到内存中，并且模块也可以在其他程序中重用。在一个模块已经被单独测试后，再与其他模块一起集成到整个程序结构中。

程序流程遵循简单的分层模型，它采用如“for”“repeat”和“while”这样的循环结构，不主张使用“Go To”语句。

结构化程序设计最早是由 Corrado Bohm 和 Guiseppe Jacopini 提议的。这两位数学家论证了任何计算机程序都可以只用 3 种结构写成：顺序、选择和循环。Edsger Dijkstra 的后续文章《GO TO 语句有害》推动了结构化程序设计的发展。最常用的模型是由 Dijkstra 开发的。在这个模型中（它通常被认为等同于结构化程序设计，虽然有其他模型存在），开发者把程序分成小节，每节只有一个入口点和一个出口点。

几乎任何语言都可以使用结构化编程技术，以避免非结构化语言的常见陷阱。非结构化编程必须依靠开发人员的纪律，以避免结构性问题，以及由此可能导致程序组织混乱。大多数现代程序语言具有鼓励结构化程序设计的特点。面向对象程序设计（OOP）可以被看作结构化程序设计的一种类型；它使用结构化程序设计技术的程序流程，并为数据模型增加了更多的结构。

3. 编译器

编译器是一个特殊的程序，用来处理用特定编程语言编写的语句，并把它们转化成计算机处理器可用的机器语言或代码。通常情况下，程序员在编辑器中用像 C 或 Pascal 这样的编程语言一次一行地编写语句，所创建的文件包含了所谓的源语句。然后程序员运行相应的语言编译器，并指定包含源语句的文件名。

在执行（运行）时，编译器首先按照语法一个接一个地解析（或分析）语句，然后，在一个或多个连续阶段或“关口”建立输出代码，确保与其他语句相关的语句在最终代码中能被正确引用。传统上，编译器的输出被称为目标代码或目标模块（注意，这里的目标——“object”与面向对象程序设计无关）。目标代码是处理器可以一次处理或执行一个指令的机器代码。

最近，Java 编程语言——在面向对象程序设计中使用的语言——引入了可以在具有 Java 虚拟机或字节码解释器平台的计算机上运行的编译输出（被称为字节码），这样可以把字节码转换为能够由实际硬件处理器执行的指令。使用此虚拟机，字节码可以有选择地在执行平台上由即时编译器重新编译。

传统上，在某些操作系统中，编译之后需要一个额外的步骤——当一个以上的目标模块要同时运行并要实现指令序列和数据的交叉引用时，需要解决指令序列或数据的相对位置的问题。这个过程有时叫作链接编辑，输出称为加载模块。

编译器与 3GL 及更高级的语言一起工作。汇编器中运行使用处理器的汇编语言编写的程序。

4. 解释器

解释器是执行由高级编程语言编写的指令的程序。有两种运行高级编程语言编写的程序的方式：最常见的是编译该程序，另一种则是通过解释器运行该程序。

解释器把高级指令转换成中间形式，然后执行它。与此相对应，编译器将高级指令直接转换成机器语言。通常编译的程序运行得比解释的程序更快。但是，解释器的优势一方面在于它不需要经过生成机器指令的编译阶段，如果程序很长，这个过程可能很费时；另一方面，解释器可以立即执行高级程序。因此，在程序的开发过程中，当程序员想一次添加小段程序并迅速对其进行测试时，有时会使用解释器。此外，解释器往往用于教学中，因为它们允许学生进行交互编程。

大多数高级编程语言都可使用解释器和编译器。然而，BASIC 和 LISP 是专门为用解释器来执行而设计的。此外，页面描述语言（如 PostScript）也使用解释器。例如，每一个 PostScript 打印机都有一个内置的执行 PostScript 指令的解释器。

5. 高级编程语言

高级编程语言与人类语言相似。与低级编程语言不同，高级编程语言对程序员友好，易于编码，调试和维护。

高级编程语言提供了对机器语言的更高层次的抽象。它们不直接与硬件交互。相反，它们更多地关注复杂的算术运算，最佳的程序效率和编码的简便性。

高级语言程序要求编译器/解释器将源代码转换为机器语言。我们可以把用高级编程语言编写的源代码编译为多种机器语言。因此，它们是机器无关的语言。

今天，几乎所有程序都是使用高级编程语言开发的。可以使用高级编程语言开发各种应用程序。高级编程语言可以用来开发桌面应用程序、网站、系统软件、实用程序软件等。

根据执行模型将高级编程语言分为两类：编译语言和解释语言。

6. 脚本语言

脚本语言是高级编程语言，由另一个程序在运行时解释，而不是由计算机的处理器编译成其他编程语言（如 C 和 C++）。脚本语言可以被嵌入到 HTML 中，通常用于给 Web 页面添加功能（如不同的菜单样式或图形显示或提供动态广告）。此类语言是客户端脚本语言，它们会影响终端用户在浏览器窗口看到的数据。其他脚本语言是服务器端脚本语言，如操作服务器上一个数据库的数据。

脚本语言的产生很大程度上是因为因特网成为一种主要的通信工具。JavaScript、ASP、JSP、PHP、Perl、Tcl 和 Python 都是脚本语言。

Unit 4

Text A

Programming Language

Text A Programming Language 

A programming language, such as C, FORTRAN or Pascal, enables a programmer to write programs that are more or less independent of a particular type of computer.[1] Such languages are considered high-level because they are closer to human languages and further from machine languages. In contrast, assembly languages are considered low-level because they are very close to machine languages.

The main advantage of high-level languages over low-level languages is that they are easier to read, write, and maintain. Ultimately, programs written in a high-level language must be translated into machine language by a compiler or interpreter.

The first high-level programming languages were designed in the 1950s. Now there are dozens of different languages, including Ada, Algol, BASIC, COBOL, C, C++, FORTRAN, LISP, Pascal, and Prolog.

1. Machine Language

The lowest-level programming language (except for computers that utilize programmable microcode). Machine languages are the only languages understood by computers. While easily understood by computers, machine languages are almost impossible for humans to use because they consist entirely of numbers.[2] Programmers, therefore, use either a high-level programming language or an assembly language. An assembly language contains the same instructions as a machine language, but the instructions and variables have names instead of being just numbers.

Programs written in high-level languages are translated into assembly language or machine language by a compiler. Assembly language programs are translated into machine language by a program called an assembler.

Every CPU has its own unique machine language. Programs must be rewritten or recompiled, therefore, to run on different types of computers.

2. Assembly Language

A programming language that is removed from a computer's machine language. Machine languages consist entirely of numbers and are almost impossible for humans to read and write. Assembly languages have the same structure and set of commands as machine languages, but they enable a programmer to use names instead of numbers.

Each type of CPU has its own machine language and assembly language, so an assembly language program written for one type of CPU won't run on another. In the early days of programming, all programs were written in assembly language. Now, most programs are written in a high-level language such as FORTRAN or C. Programmers still use assembly language when speed is essential or when they need to perform an operation that isn't possible in a high-level language.

3. BASIC

Acronym for Beginner's All-purpose Symbolic Instruction Code. Developed by John Kemeney and Thomas Kurtz in the mid 1960s at Dartmouth College, BASIC is one of the earliest and simplest high-level programming languages. During the 1970s, it was the principal programming language taught to students, and continues to be a popular choice among educators.

Despite its simplicity, BASIC is used for a wide variety of business applications. There is an ANSI standard for the BASIC language, but most versions of BASIC include many proprietary extensions. Microsoft's popular Visual Basic, for example, adds many object-oriented features to the standard BASIC.

Recently, many variations of BASIC have appeared as programming, or macro, languages within applications. For example, Microsoft Word and Excel both come with a version of BASIC with which users can write programs to customize and automate these applications.[3]

4. COBOL

Acronym for common business-oriented language. Developed in the late 1950s and early 1960s, COBOL is the second-oldest high-level programming language (FORTRAN is the oldest). It is particularly popular for business applications that run on large computers.

COBOL is a wordy language; programs written in COBOL tend to be much longer than the same programs written in other languages. This can be annoying when you program in COBOL, but the wordiness makes it easy to understand programs because everything is spelled out. Although disparaged by many programmers for being outdated, COBOL is still the most widely used programming language in the world.

5. FORTRAN

Acronym for formula translator, FORTRAN is the oldest high-level programming language. Designed by John Backus for IBM in the late 1950s, it is still popular today, particularly for scientific applications that require extensive mathematical computations.

The two most common versions of FORTRAN are FORTRAN IV and FORTRAN 77. FORTRAN IV was approved as a USASI standard in 1966. FORTRAN 77 is a version of FORTRAN that was approved by ANSI in 1978 (they had expected to approve it in 1977, hence the name). FORTRAN 77 includes a number of features not available in older versions of FORTRAN. A new ISO and ANSI standard for FORTRAN, called FORTRAN 90, was developed in the early 1990s.

6. C

A high-level programming language developed by Dennis Ritchie at Bell Labs in the mid 1970s. Although originally designed as a systems programming language, C has proved to be a powerful and flexible language that can be used for a variety of applications, from business programs to engineering.[4] C is a particularly popular language for personal computer programmers because it is relatively small — it requires less memory than other languages.

The first major program written in C was the UNIX operating system, and for many years C was considered to be inextricably linked with UNIX. Now, however, C is an important language independent of UNIX.

Although it is a high-level language, C is much closer to assembly language than most other high-level languages. This closeness to the underlying machine language allows C programmers to write very efficient code. The low-level nature of C, however, can make the language difficult to use for some types of applications.

7. Python

7.1 Features of Python

7.1.1 Code Quality

Python code is highly readable, which makes it more reusable and maintainable.[5] It has broad support for advanced software engineering paradigms such as object-oriented (OO) and functional programming.

7.1.2 Developer Productivity

Python has a clean and elegant coding style. It uses an English-like syntax and is dynamically-typed. So, you never declare a variable. A simple assignment binds a name to an object of any type. Python code is significantly smaller than the equivalent C++/Java code. It implies there is less to type, limited to debug, and fewer to maintain. Unlike compiled languages, Python programs don't need to compile and link which further boosts the developer speed.

7.1.3 Code Portability

Since Python is an interpreted language, so the interpreter has to manage the task of portability. Also, Python's interpreter is smart enough to execute your program on different platforms to produce the same output. So, you never need to change a line in your code.

7.1.4 Built-in and External Libraries

Python packages a large number of the prebuilt and portable set of libraries. You can load them when needed to use the desired functionality.

7.1.5 Component Integration

Some applications require interaction across different components to support the end to end workflows. One such component could be a Python script while others be a program written in languages like Java/C++ or any other technology.

Python has several ways to support the cross-application communication. It allows mechanisms like loading of C and C++ libraries or vice-versa, integration with Java and DotNET(.NET) components, communication using COM/Silverlight, and interfacing with USB devices over serial ports. It can even exchange data over networks using protocols.

7.1.6 Free to Use, Modify and Redistribute

Python is an OSS. You are free to use it, make amends in the source code and redistribute, even for commercial interests. It is because of such openness that Python has garnered a vast community base which is continually growing and adding value.[6]

7.1.7 Object-Oriented from the Core

Python primarily follows the object-oriented programming (OOP) design. OOP provides an intuitive way of structuring your code, and a solid understanding of the concepts behind it can let you make the most out of your coding.

7.2 Python Programming Domains

7.2.1 Web Application Development

Python has the lion's share in the field of Web development. Many employers look for full-stack programmers who know Python. And you can become one of them by learning frameworks like Django, Flask, CherryPy, and Bottle, which give extensive support for web development.

7.2.2 Scientific and Numeric Computing

Python has become the obvious choice for working in scientific and numeric applications. And there are multiple reasons for this advancement. First and foremost is that Python is a free and open source language. And it allows to modify and redistribute its source code.

Next, the reason for becoming it more dominant in the field of scientific and Numeric is the rapidly growing number of specialized modules. All of these are available for free.

Hence, Python is becoming a leader in this field. The focus of Python language is to bring more productivity and increase readability.

7.2.3 GUI Programming

Python has some inherent qualities like clean and straightforward coding syntax as well as dynamic typing support. These work as the catalyst while developing complex GUI and image processing applications.

Python's clean syntax and tremendous support of many GUI libraries made Programmers deliver graphics software.

7.2.4 Software Prototyping

Python has many qualities that make it a natural choice for prototyping. The first is being an open source programming language, a massive number of users follow and contribute to its development. Further, the lightness, versatility, scalability, and flexibility of refactoring code in Python speed up the development process from the initial prototype. Hence, Python gives you an easy-to-use interface to create prototypes.

7.2.5 Professional Training

Python is indeed the right programming language for teaching and training purposes. It can be a stepping stone for beginners to enter into professional training. They can even cover overlapping areas like data analytics and machine learning.

Hence, there is a huge demand for professional trainers who can teach both basic and advanced level Python programming. You can impart training offline in a classroom or use tools to do it online.

8. R Programming Language

R has become a very popular programming language and development environment used for statistical computation and graphics. It started off as a GNU project that was similar to the language S but had many other features. It can be used to compute a wide variety of statistical tests that include the classic tests like student's test and correlation test. It has a powerful user base and strong community support. It is freely available to all.

8.1 What is the R programming Language?

The R programming language consists of a huge variety of statistical and graphical methods. It contains regression analysis algorithms, machine learning, time series and many more. Most of its packages or libraries are written in R.

To increase its efficiency, the procedures written in C, C++, FORTRAN, Python, and .NET can be integrated. The language has become popular among academic institutions as well as large corporations such as Uber, Google, and Facebook.

The core of the language is actually an interpreted programming language that supports modular programming, looping, and branching. It is heavily used in data analysis that is performed through the following steps:

- Programming.
- Transforming.
- Discovering.
- Modelling.
- Communicating the results.

The R language programming environment is based on a command line interface.

8.2 Features of R language

The different features of the R programming language are as follows:

- It is simple and effective that contains conditional loops, recursive functions, and input/output facilities.
- Its functions support procedural programming. Object-oriented programming is supported by the generic functions.
- Matrix arithmetic operations are supported by the language.
- Being an interpreted language, it can be operated through the command line.
- Storage and data handling provisions are available.
- It has many operations that can be used to work on arrays, lists, matrices, and vectors.
- Features such as exporting data, handling database input, data viewing, variable labels are available.
- It provides facilities for displaying the results of graphical analysis, as graphs either on screen or as a hardcopy.
- R code can be run directly in the console without a compiler.

New Words

programming	['prəʊgræmɪŋ]	*n.*编程
programmer	['prəʊgræmə]	*n.*程序员

compiler	[kəm'paɪlə]	*n.*编译器，编译程序
programmable	['prəʊgræməbl]	*adj.*可编程的，可设计的
microcode	['maɪkrəʊkəʊd]	*n.*微代码，微码
impossible	[ɪm'pɒsəbl]	*adj.*不可能的，不会发生的，难以忍受的
variable	['veərɪəbl]	*n.*变量
assembler	[ə'semblə]	*n.*汇编程序
unique	[ju: 'ni:k]	*adj.*唯一的，独特的
principal	['prɪnsɪpəl]	*adj.*主要的，首要的
object-oriented	['əbdʒekt'ɔ:rɪəntɪd]	*adj.*面向对象的
feature	['fi:tʃə]	*n.*特征，特点
macro	['mækrəʊ]	*n.*宏
customize	['kʌstəmaɪz]	*v.*定制，用户化
wordy	['wɜ:dɪ]	*adj.*多言的，冗长的，废话连篇的，唠叨的
annoying	[ə'nɔɪɪŋ]	*adj.*恼人的，讨厌的
wordiness	['wɜ:dɪnɪs]	*n.*累赘，冗长
disparage	[dɪ'spærɪdʒ]	*vt.*贬低，毁谤，轻视
outdated	[ˌaʊt'deɪtɪd]	*adj.*过时的，不流行的
computation	[ˌkɒmpju'teɪʃn]	*n.*计算，估计
approve	[ə'pru:v]	*v.*批准，认可，通过
powerful	['paʊəfʊl]	*adj.*强大的，有力的
inextricably	[ˌɪnɪk'strɪkəbli]	*adv.*逃不掉地，解决不了地，解不开地
quality	['kwɒlətɪ]	*n.*质量，品质
readable	['ri:dəbl]	*adj.*可读的，易读的；易懂的
reusable	[ˌri: 'ju:zəbl]	*adj.*可再用的，可重复使用的
maintainable	[meɪn'teɪnəbl]	*adj.*可维护的
functional	['fʌŋkʃənl]	*adj.*功能的；函数的
productivity	[ˌprɒdʌk'tɪvɪtɪ]	*n.*生产率，生产力
elegant	['elɪgənt]	*adj.*简练的，简洁的；漂亮的
dynamically	[daɪ'næmɪklɪ]	*adv.*动态地
declare	[dɪ'kleə]	*vt.*声明
assignment	[ə'saɪnmənt]	*n.*赋值，分配
bind	[baɪnd]	*vt.*绑定
equivalent	[ɪ'kwɪvələnt]	*adj.*相等的，相当的，等效的
link	[lɪŋk]	*n.* & *v.*链接
library	['laɪbrərɪ]	*n.*库
prebuilt	[pre'bɪlt]	*adj.*预建的，预制的
component	[kəm'pəʊnənt]	*n.*部件；成分
integration	[ˌɪntɪ'greɪʃn]	*n.*结合；整合；一体化
workflow	['wɜ:kfləʊ]	*n.*工作流程

redistribute	[ˌri:dɪ'strɪbju:t]	*vt.*重现发布；重新分配
amend	[ə'mend]	*v.*改良，修改；修订
continually	[kən'tɪnjuəlɪ]	*adv.*不停地；持续地；屡屡地；一再地
intuitive	[ɪn'tju:ɪtɪv]	*adj.*直觉的；直观的
correlate	['kɒrəleɪt]	*v.*使互相关联；联系 *adj.*相关的
share	[ʃeə]	*v.*共享，分享
full-stack	[fʊl-stæk]	*adj.*全栈的；全能的；完整的
deliver	[dɪ'lɪvə]	*vt.*发表；交付
essential	[ɪ'senʃəl]	*adj.*基本的；必要的；本质的
task	[tɑ:sk]	*n.*工作，任务；作业
obvious	['ɒbvɪəs]	*adj.*明显的；显著的
dominant	['dɒmɪnənt]	*adj.*占优势的；统治的，支配的
readability	[ˌri:də'bɪlətɪ]	*n.*可读性
inherent	[ɪn'hɪərənt]	*adj.*固有的，内在的
straightforward	[ˌstreɪt'fɔ:wəd]	*adj.*直截了当的；坦率的；明确的 *adv.*直截了当地；坦率地
tremendous	[trə'mendəs]	*adj.*极大的，巨大的；极好的
prototype	['prəʊtətaɪp]	*n.*原型
versatility	[ˌvɜ:sə'tɪlətɪ]	*n.*多用途
statistical	[stə'tɪstɪkl]	*adj.*统计的，统计学的
regression	[rɪ'greʃn]	*n.*（统计学）回归
modular	['mɒdjʊlə]	*adj.*模块化的
loop	[lu:p]	*n.*循环
branch	[brɑ:ntʃ]	*n.*分支
transform	[træns'fɔ:m]	*vt.*改变；改观；变换 *vi.*改变
recursive	[rɪ'kɜ:sɪv]	*adj.*回归的，递归的
array	[ə'reɪ]	*n.*数组
hardcopy	['ha:d'kɒpɪ]	*n.*硬拷贝
console	[kən'səʊl]	*n.*控制台，操纵台

Phrases

be independent of	无关，不依赖，不取决于，不受……限制或制约
more or less	或多或少
assembly language	汇编语言
high-level language	高级编程语言
translate … into …	把……翻译为……
instead of	代替，而不是……

spell out	讲清楚，清楚地说明
Bell Labs	贝尔实验室
software engineering	软件工程
coding style	编程风格，编码风格
code portability	代码可移植性
interpreted language	解释语言
source code	源编码，源程序
development environment	开发环境
command line interface	命令行接口
conditional loop	条件循环
recursive function	递归函数

Abbreviations

ANSI (American National Standards Institute)	美国国家标准协会
USASI (United States of America Standards Institute)	美国标准学会
ISO (International Organization for Standardization)	国际标准化组织
OO (Object-Oriented)	面向对象的
COM (Component Object Model)	组件对象模型
USB (Universal Serial Bus)	通用串行总线
OSS (Object Storage Service)	对象存储服务
GNU	"GNU is Not UNIX"的递归缩写

Notes

[1] A programming language, such as C, FORTRAN or Pascal, enables a programmer to write programs that are more or less independent of a particular type of computer.

本句中，that are more or less independent of a particular type of computer是一个定语从句，修饰和限定programs。enables sb. to do sth.的意思是"使某人能够做某事"，be independent of的意思是"与……无关，不依赖，不取决于，不受……限制或制约"，more or less 的意思是"或多或少"。请看下例：

The consumer protection law enables any impaired consumer to claim money from the company.

消费者权益保护法使任何受损害的消费者都有向公司索赔的权利。

Young people should be independent of their parents.

年轻人应该不依赖父母，要自立。

We hope our explanation will prove more or less helpful.

希望我们的说明或多或少有些帮助。

[2] While easily understood by computers, machine languages are almost impossible for humans to use because they consist entirely of numbers.

本句中，While 的意思是"尽管，虽然"，easily understood by computers 是个过去分词短

语，修饰 machine languages，While easily understood by computers 可以扩展为一个让步状语从句：While machine languages are easily understood by computers。

[3] For example, Microsoft Word and Excel both come with a version of BASIC with which users can write programs to customize and automate these applications.

本句中，with which users can write programs to customize and automate these applications 是一个介词前置的定语从句，修饰和限定 a version of BASIC。

[4] Although originally designed as a systems programming language, C has proved to be a powerful and flexible language that can be used for a variety of applications, from business programs to engineering.

本句中，Although originally designed as a systems programming language是一个让步状语从句，修饰谓语 has proved。that can be used for a variety of applications, from business programs to engineering 是一个定语从句，修饰和限定 a powerful and flexible language。

[5] Python code is highly readable, which makes it more reusable and maintainable.

本句中，which makes it more reusable and maintainable 是一个非限定性定语从句，对主句 Python code is highly readable 进一步补充说明。在该从句中，which 指它前面的整个句子，it 指 Python code。

[6] It is because of such openness that Python has garnered a vast community base which is continually growing and adding value.

本句使用了强调 it 句型：It is/was +被强调部分+that...，强调 Python has garnered a vast community base 的原因。

本句中，which is continually growing and adding value 是一个定语从句，修饰和限定 a vast community base。

Text B

Computer Programmer

1. What is a Computer Programmer?

Text B Computer Programmer

A computer programmer, or coder, is someone who writes computer software. The term computer programmer can refer to a specialist in one area of computer programming or to a generalist who writes code for many kinds of software.

One who practices or professes a formal approach to programming may also be known as a programmer analyst. The term programmer can be used to refer to a software developer, software engineer, computer scientist, or software analyst. However, members of these professions typically possess other software engineering skills beyond programming. For this reason, the term programmer is sometimes considered an oversimplification of these other professions.

This has sparked much debate among developers, analysts, computer scientists, programmers, and outsiders who continue to be puzzled at the subtle differences in these occupations. Within software engineering, programming (the implementation) is regarded as one phase in a software developmental process.

2. What Does a Computer Programmer Do?

The 21st century has brought in an extraordinary amount of technological progress. In the centre of this modern technology sits computer programmers, with the technological skills to create and navigate any new projects that may come their way. It's the job of computer programmers to take designs created by software developers and engineers and turn them into sets of instructions that computers can follow. These instructions result in the social media platforms, word processing programs, browsers, and more that people use every day.

There is an ongoing debate on the extent to which the writing of programs is an art, a craft, or an engineering discipline. In general, good programming is considered to be the measured application of all three, with the goal of producing an efficient software solution.

A computer programmer figures out the process of designing, writing, testing, debugging/troubleshooting and maintaining the source code of computer programs. This source code is written in a programming language so the computer can "understand" it. The code may be a modification of an existing source or something completely new.

The purpose of programming is to create a program that produces a certain desired behaviour (customization). The process of writing source code often requires expertise in many different subjects, including knowledge of the application domain, specialized algorithms, and formal logic.

The computer programmer also designs a graphical user interface (GUI) so that non-technical users can use the software through easy, point-and-click menu options. The GUI acts as a translator between the user and the software code.

Some, especially those working on large projects that involve many computer programmers, use computer-assisted software engineering (CASE) tools to automate much of the coding process. These tools enable a programmer to concentrate on writing the unique parts of a program. A programmer working on smaller projects will often use "programmer environments" or applications that increase productivity by combining compiling, code walk-through, code generation, test data generation, and debugging functions.

A computer programmer will also use libraries of basic code that can be modified or customized for a specific application. This approach yields more reliable and consistent programs and increases programmers' productivity by eliminating some routine steps. The computer programmer will also be responsible for maintaining the program's health.

As software design has continued to advance, and some programming functions have become automated, computer programmers have begun to assume some of the responsibilities that were once performed only by software engineers. As a result, some computer programmers now assist software engineers in identifying user needs and designing certain parts of computer programs, as well as other functions.

3. What Types of Computer Programmers are There?

The products we use everyday, such as our computer, our smart phone, and our car, wouldn't be able to do what we ask them to do if it weren't for computer programmers. Computer programming is a very detail-oriented profession. Therefore, programmers are required to focus on code for long periods of time without losing focus or losing track of their progress. Often small but critical code issues can have a big impact technically, and can prevent a program from operating correctly. Persistence and the ability to detect and rectify small discrepancies as quickly as possible are essential in order to solve issues.

There are four main categories of computer programmers. The following describes what the differences are between them and their roles.

3.1 Computer Hardware Programmer

Computers have their own machine language that they are able to understand and take instructions from. Computer hardware programmers write these instructions in a specific machine language (code) so that a computer knows what to do when someone presses the power button, types on the keyboard, or uses the mouse. They also write code so that text can be displayed when the computer turns on. Computer programs (a collection of instructions) are put in permanent memory storage so that as soon as someone turns on their computer, programmed information is displayed that prompts the user to choose what the computer does next.

Computer hardware programmers are also involved in researching, designing, developing, and testing computer equipment.

3.2 Web Developer

Computer programmers that design, create, and modify the millions of websites found on the Internet are called Web developers. These types of computer programmers use software that allows them the ability to dictate what kinds of functions people are able to do when they access a website.

Web developers start by analyzing a user's needs before designing and structuring a website. They also add applicable graphics, audio, and video components if needed (often using software designed specifically to enable the creation of Web and multimedia content). Even a simple blog needs a Web developer that can design the structure, function and the information that an audience can see.

Not only are Web developers responsible for the look of a website, but they are also responsible for its performance, capacity, and sometimes content creation as well.

3.3 Software Developer

Software developers are computer programmers that focus on designing and managing programming functions. A function is a section of organized, reusable code that is used to perform an action (functions can also be called methods, sub-routines, and procedures).

It is possible for software developers to build entire software applications with only functions. These functions, for example, can enable a person to open their tax file, edit their tax documents, and then save or print them. Each type of program is designed differently and has instructions and specific tasks relevant to the company it services. So trying to edit photos in your bank software won't work because the software developer's design doesn't include any instructions for your bank program to edit photos. Software developers also develop, design, create, and modify programs that run the operating systems for computers, networks and even smart phones. In a nutshell, a software developer's goal is to optimize operational efficiency by designing customized software.

3.4 Database Developers

A database (an organized collection of data) collects, arranges, sorts and retrieves related pieces of information. It runs behind the scenes of user software and websites, and is generally stored and accessed electronically from a computer system.

Database developers (or database programmers) are the people who are responsible for creating and implementing computer databases. These types of computer programmers will analyze the data needs of a company and then produce an effective database system to meet those needs. They will also test database programs for efficiency and performance, troubleshoot and correct any problems that come up. Governments, banks, car dealerships, ecommerce businesses (think Amazon) all need specialized business systems and rely on their databases and the database developers who customize and revise them.

New Words

coder	['kəʊdə]	*n.*写代码的人，编码者
generalist	['dʒenrəlɪst]	*n.*通才，多面手
profess	[prə'fes]	*vt.*声称；宣称；公开表明；信奉
analyst	['ænəlɪst]	*n.*分析师
possess	[pə'zes]	*vt.*拥有；掌握，懂得
oversimplification	['əʊvəˌsɪmplɪfɪ'keɪʃn]	*n.*过度简化（的事物）
debate	[dɪ'beɪt]	*n.*讨论；辩论；争论
puzzle	['pʌzl]	*vt.*使迷惑，使难解 *n.*难题
subtle	['sʌtl]	*adj.*微妙的；敏感的
occupation	[ɒkju: 'peɪʃn]	*n.*职业，工作
phase	[feɪz]	*n.*阶段
extraordinary	[ɪk'strɔ:dɪnərɪ]	*adj.*非凡的；特别的
browser	['braʊzə]	*n.*浏览器，浏览程序
craft	[krɑ:ft]	*n.*手艺 *vt.*手工制作；精巧地制作
debug	[di: 'bʌg]	*vt.*调试程序，排除故障
troubleshoot	['trʌblʃu:t]	*v.*故障排除，故障检测
customization	['kʌstəmaɪzeɪʃn]	*n.*用户化，专用化，定制
menu	['menju:]	*n.*菜单
option	['ɒpʃn]	*n.*选项
concentrate	['kɒns əntreɪt]	*v.*专心于；注意；集中
detail-oriented	['di:teɪl-'ɔ:rɪəntɪd]	*adj.*细节导向的，面向细节的，注重细节的
persistence	[pə'sɪstəns]	*n.*坚持不懈；执意；持续

discrepancy	[dɪs'krepənsi]	*n.*矛盾；不符合（之处）
permanent	['pɜ:mənənt]	*adj.*永久（性）的，不变的，持久的
prompt	[prɒmpt]	*v.*提示
website	['websaɪt]	*n.*网站
dictate	[dɪk'teɪt]	*vt.* 命令，指示；控制，支配；
multimedia	[ˌmʌltɪ'mi:dɪə]	*n.*多媒体 *adj.*多媒体的
blog	[blɒg]	*n.*博客
capacity	[kə'pæsɪtɪ]	*n.*容量
subroutine	['sʌbru:ti:n]	*n.*子例程
procedure	[prə'si:dʒə]	*n.*过程

Phrases

be known as	被称为
software developer	软件开发人员
software engineer	软件工程师
be regarded as	被认为是
technological progress	技术进步
word processing program	字处理程序
formal logic	形式逻辑
code walk-through	代码走查
code generation	代码生成
test data generation	测试数据生成
routine step	常规步骤
critical code	关键代码
Web developer	Web 开发人员

Abbreviations

CASE (Computer-Aided Software Engineering)　　计算机辅助软件工程

Word Building

在英语中，一些单词可以从一种词类转换为另一种词类，这叫作“转化”。转化后的词义往往与原来的词义有密切的联系。转化的方法主要有以下几种。

1. 名词转化为动词

machine 机器 —— to machine 机加工

time 时间 —— to time 计时，定时

format 格式 —— to format 格式化

2. 动词转化为名词

to talk 交谈 —— talk 谈话，讲话

to test 测验，检查 —— test 测验，检验

to use 使用 —— use 用途（注意发音不同）

to increase 增加 —— increase 增加，增量（注意重音不同）

3. 形容词转化为名词

mineral 矿物的 —— mineral 矿物质

good 好的 —— good 益处

final 最后的 —— final 决赛

4. 其他词转化为名词

twos and threes 三三两两

a must 必要条件

注意，一些以辅音结尾的单词转换词类后,词尾辅音的读音会发生变化，有的拼写也会发生变化。一些双音节词转换词类后，单词的重音也会发生变化。名词的重读音节在第一个音节，而动词的重读音节在第二个音节。

在计算机英语中，单词的转化尤以名词与动词之间的转化为多。

Career Training

面试技巧

现在，许多公司都通过面试来录用工作人员，它们希望通过面试来了解求职者能否胜任工作。因此，成功的面试对求职者尤为重要。

求职者要得到一份称心如意的工作，除了必要的知识和经验之外，还需要具备一定的面试技巧。

面试前，求职者应该了解应聘公司的一些情况，如该公司的组织结构、规模、公司的产品和服务、所感兴趣的部门、工作描述、员工的福利待遇及公司的地理位置等。同时还应该为一些面试中常问的问题准备好答案。准备面试时的服饰（衣着要大方得体）。提前 15 分钟到面试地点，对接待员或秘书要友好，他们的意见有时也很重要。

面试中，求职者要有信心、面带微笑，坐在面试人员的对面，要坐直、放松，不要紧张。认真听面试人员的问题，如果问题很长，可以做一些笔记。在回答问题时要表述清晰、语调友好，不要只回答“是”或“不是”，要阐明和详述自己的观点，但不要漫无边际。

面试中，求职者要积极热情，也可以提一些问题，如：What are the company's greatest strengths? In what areas is the company trying to improve? How much travel is involved? What kind of assignments could I expect in the first 6 months? What qualities are you looking for in a candidate? 求职者也可以问面试人员什么时间可以得到答复。

面试结束时一定要向面试人员致谢。

面试中常问的问题如下：

1. What extracurricular activities were you involved in? What made you choose those? Which

of them did you enjoy most, and why?

2. What led you to select your major? You minor?

3. Which of your courses did you like the least?

4. Was there a course you found particularly challenging?

5. If you were to start college over again tomorrow, what are the courses you would take? Why?

6. In college, how did you go about influencing someone to accept your ideas?

7. Based on what you know of the job market, which of your courses were the most useful?

8. What advice would you give college student intending to go into your field?

9. What are your most memorable experiences from college?

10. What did you learn from your internships or work experiences?

11. Why don't I see internships or work-study experiences on you resume?

12. In what courses did your worst grades? Why? How do you think that will affect your performance on the job?

13. Why did you decide to go to college?

14. How was your college education funded?

15. What percentage of your college did you college did you pay for and what sort of jobs did you have while you were in school?

16. Tell me a little about some of your extracurricular activities that would assist you in this job.

17. Why are you working in a field other than the one in which you have a degree?

18. What have you done to stay current in your field?

19. Are you satisfied with the grades you received in school?

20. Do you think your grades accurately reflect your ability?

软件水平考试试题解析

【真题再现】

从供选择的答案中选出应填入下列英语文句中____内的正确答案，把编号写在答卷的对应栏内。

The use of the computer is changing the very __A__ of many jobs that exist within a business. In the industrial __B__, tools were developed to assist in improving production, but much work still involved __C__ labor. The information __B__ has brought about another change——a change from __C__ labor to __D__ labor.

Pressure on computer worker can be great. Whether operating a robot, running a computer, or programming a computer, a single error can be __E__. The smallest error could misdirect an airline, disrupt delivery schedules, or cost millions of dollars.

供选择的答案

A～E：①important ②great ③disastrous ④physical ⑤body

⑥brain ⑦mental ⑧revolving ⑨revolution ⑩nature

【答案】A：⑩　B：⑨　C：④　D：⑦　E：③

【试题解析】

A：根据句意，此处应该填一个名词，意思是“本质”，the nature of many jobs 的意思是“许多工作的本质”。故选⑩。

B：In the industrial revolution 的意思是“在工业革命中”，故选⑨。

C：在工业革命中，许多工作仍是体力劳动（physical labor），而不是脑力劳动（mental labor），故选④。

D：The information revolution 的意思是“信息革命”，它与工业革命不同，人们的工作性质发生了变化，即从体力劳动转变为脑力劳动。故选⑦。

E：根据句意，此处应该填一个形容词，disastrous 的意思是“损失惨重的，灾难性的”。故选③。

【参考译文】

计算机的使用正在改变许多工作的本质。虽然在工业革命中，研制工具以改进产品，但是许多工作还是体力劳动。信息革命带来了另一场变革——从体力劳动到脑力劳动的变革。

计算机工作人员的压力很大。无论是操作机器人、运行计算机还是给计算机编程，一个错误就可能导致一场灾难。即便是最小的错误都可能误导航线、打乱交货时间表或导致数百万美元的损失。

Exercises

[Ex. 1] 根据 Text A 回答以下问题。

1) Why are such languages as C, FORTRAN or Pascal are considered high-level?

2) Why are assembly languages considered low-level?

3) What is the main advantage of high-level languages over low-level languages?

4) What are programs written in high-level languages translated into? How? And what about assembly language programs?

5) When do programmers still use assembly language?

6) What is FORTRAN? Who designed it? When? What is it particularly popular for?

7) Why is C a particularly popular language for personal computer programmers?

8) What are the features of Python mentioned in the passage?

9) What are the Python programming domains mentioned in the passage?

10) What is the core of the R language?

[Ex. 2] 根据 Text B 回答以下问题。

1) What can the term computer programmer refer to?

2) What is programming (the implementation) regarded as within software engineering?

3) In general, good programming is considered to be the measured application of all three, with the goal of producing an efficient software solution.

4) What does a computer programmer figure out?

5) What does the process of writing source code often require?

6) What have computer programmers begun to do as software design has continued to advance, and some programming functions have become automated?

7) How many main categories of computer programmers are mentioned in the passage? What are they?

8) What are Web developers responsible for?

9) What are software developers?

10) What are database developers?

[Ex. 3] 把下列句子翻译为中文。

1) Computer programs can be written in a variety of different languages.

2) Common languages include Java, C, C++, FORTRAN, Pascal, LISP, and BASIC.

3) Some people classify languages into two categories, higher-level and lower-level.

4) In general, higher-level languages can be either interpreted or compiled.

5) The program that converts assembly language to machine language is called an assembler.

6) Each processor's machine language will be different from other processors' machine

language.

7) C is one of the most popular computer languages in the world.

8) R is a programming language that is primarily used for statistical computing and graphics. It is available for free.

9) Python is a programming language that lets you work quickly and integrate systems more effectively.

10) Many people do not consider machine language and assembly language at all when talking about programming languages.

[Ex. 4] 软件水平考试真题自测。

从供选择的答案中选出应填入下列英语文句中____内的正确答案，把编号写在答卷的对应栏内。

The beauty of software is in its function, in its internal structure, and in the way in which it is created by a team.To a user, a program with just the right features presented through an intuitive and ____1____ interface is beautiful. To a software designer, an internal structure that is partitioned in a simple and intuitive manner, and that minimizes internal coupling is beautiful. To developers and managers, a motivated team of developers making significant progress every week, and producing defect-free code, is beautiful. There is beauty on all these levels. Our world needs software—lots of software. Fifty years ago software was something that ran in a few big and expensive machines. Thirty years ago it was something that ran in most companies and industrial settings. Now there is software running in our cell phones, watches, appliances, automobiles, toys, and tools. And the need for new and better software never ____2____. As our civilization grows and expands, as developing nations build their infrastructures, as developed nations strive to achieve ever greater efficiency, the need for more and more software ____3____ to increase. It would be a great shame if, in all that software, there was no beauty. We know that software can be ugly. We know that it can be hard to use, unreliable, and carelessly structured. We know that there are software systems whose tangled and careless internal structures make them expensive and difficult to change. We know that there are software systems that present their features through an awkward and cumbersome interface. We know that there are software systems that crash and misbehave. These are ____4____ systems. Unfortunately, as a profession, software developers tend to create more ugly systems than beautiful ones. There is a secret that the best software developers know. Beauty is cheaper than ugliness. Beauty is faster than ugliness. A beautiful software system can be built and maintained in less time, and for less money, than an ugly one. Novice software developers don't understand this. They think that they have to do everything fast and quick. They think that beauty is ____5____. No! By doing things fast and quick, they make messes that make the software stiff, and hard to understand. Beautiful systems are flexible and easy to understand. Building them and maintaining them is a joy. It is ugliness that is impractical. Ugliness will slow you down and make your software expensive and brittle. Beautiful systems cost the least build and maintain, and are delivered soonest.

1) A. simple	B. hard	C. complex	D. duplicated
2) A. happens	B. exists	C. stops	D. starts
3) A. starts	B. continues	C. appears	D. stops
4) A. practical	B. useful	C. beautiful	D. ugly
5) A. impractical	B. perfect	C. time-wasting	D. practical

[Ex. 5] 听短文填空。

Exercises 4

C is a programming language developed by Dennis Ritchie at Bell Laboratories in 1972. It is so named because its immediate predecessor was the B ____1____ language. Although C is____2____ by many people to be more a machine-independent ____3____ language than a ____4____ language, its close ____5____ with the UNIX ____6____ system, its enormous popularity, and its standardization by the American National Standards Institute (ANSI) have made it perhaps the closest thing to a____7____ programming language in the microcomputer/workstation marketplace. C is a compiled language that contains a small set of built-in ____8____ that are machine dependent. The rest of the C functions are machine independent and are contained in libraries that can be ____9____ from C programs. C programs are composed of one or more functions ____10____ by the programmer; thus C is a structured programming language.

Reference Translation

编 程 语 言

编程语言，例如C、FORTRAN或者Pascal，可以让程序员写出或多或少不依赖某种计算机的程序。这些语言被认为是高级编程语言，因为它们更接近人类语言且远离机器语言。相反，汇编语言被认为是低级语言，因为它非常接近机器语言。

高级编程语言比低级语言优越，主要在于它们易读、易写和易维护。高级编程语言写出的程序最终必须由编译程序和解释程序翻译成机器语言。

第一个高级编程语言设计于 20 世纪 50 年代。现在有很多种这类语言，包括Ada，Algol，BASIC，COBOL，C，C++，FORTRAN，LISP，Pascal，以及Prolog。

1. 机器语言

它是等级最低的编程语言（除了用可编程的微码的计算机）。机器语言是计算机可以理解的唯一语言。尽管可以容易地被计算机理解，但机器语言几乎不可能被人类所用，因为它们包含的全是数字。因此程序员使用高级编程语言或者汇编语言。汇编语言包含和机器语言一样的指令，但是指令和变量有自己的名字，而不仅仅是数字。

用高级编程语言编写的程序通过编译程序翻译成汇编语言或机器语言。用汇编语言所写的程序通过汇编程序翻译成机器语言。

每个中央处理器都有它唯一的机器语言。程序必须被重写或编译，才能在不同种类的计算机上运行。

2. 汇编语言

一种脱离计算机机器语言的编程语言。机器语言全部由数字组成，几乎不可能被人类读写。汇编语言与机器语言有同样的结构和命令集，但是它允许程序员运用名称代替数字。

每种中央处理器都有它独特的机器语言和汇编语言，因此，为一种中央处理器编写的汇编语言程序不能运行在其他中央处理器上。早期的编程中，所有程序都用汇编语言编写。现在大多数程序用高级编程语言，如使用FORTRAN 或 C 编写。在追求速度或者需要完成一项用高级编程语言无法完成的任务时，程序员仍会使用汇编语言。

3. BASIC

初学者通用指令码的首字母缩写。BASIC 由 John Kemeney 和 Thomas Kurtz 于 20 世纪 60 年代在 Dartmouth 大学开发，是最早、最简单的高级编程语言之一。在 20 世纪 70 年代，它是最主要的教给学生的编程语言，现在仍然是教师的热门选择。

尽管十分简单，BASIC 在商业应用方面也应用广泛。BASIC 语言有一个 ANSI 标准，但大多数 BASIC 版本包含许多专门的扩展。例如，流行的Microsoft公司的 Visual Basic，就在标准 BASIC 上增加了许多面向对象的特色。

最近，有许多以程序设计、宏或者应用软件内嵌语言形式出现的变种 BASIC。例如，Microsoft Word 和 Excel 都来自一个版本的 BASIC，在该版本中用户可以编写程序来定制、自动操作这些应用软件。

4. COBOL

COBOL 是面向商业的通用语言的首字母缩写，开发于 20 世纪 50 年代末至 60 年代初，它是第二古老的高级编程语言（第一是FORTRAN），主要用于大型计算机上的商用软件。

COBOL 是一个冗长的语言，同样一个程序，用 COBOL 编写的总是比用其他语言编写的要长得多。也许用 COBOL 编程时很恼人，但其冗长的特点却让程序更易懂，因为所有东西都已被讲清楚。尽管被许多程序员贬低为过时，但 COBOL 仍是世界上使用最广泛的编程语言。

5. FORTRAN

FORTRAN 是公式翻译程序语言的缩写，是最老的高级编程语言。它于 20 世纪 50 年代末由 John Backus 为IBM开发，但现在仍然很流行，尤其是在需要大量数学运算的科学应用软件方面。

FORTRAN 最常见的两个版本是 FORTRAN IV 和 FORTRAN 77。FORTRAN IV 于 1966 年被批准为 USASI 标准。FORTRAN 77 是由 ANSI 在 1978 年批准的版本（曾预期在 1977 年批准它，因此得名）。FORTRAN 77 包括很多旧版本没有的特点。20 世纪 90 年代初，一个ISO和 ANSI 标准的、名为 FORTRAN 90 的 FORTRAN 版本被开发出来。

6. C

C 是由 Dennis Ritchie 于 20 世纪 70 年代开发于贝尔实验室的一种高级编程语言。尽管最初被开发为系统编程语言，但 C 证明了自己是很强大且灵活的语言，可用于从商业程序到工程的各种应用程序。C 很受个人计算机程序员欢迎，因为它相对小巧——需要的内存比其他语言都少。

第一个用 C 编写的主流程序是UNIX操作系统，在很多年里，C 被认为无法解开和 UNIX 的联系。然而，现在 C 是一款独立于 UNIX 的重要语言。

尽管 C 是高级编程语言，但比其他同类都更接近汇编语言。这点让 C 程序员可以写出非常有效的代码。然而 C 所具有的低级语言属性使其难以用于某些种类的应用程序。

7. Python

7.1 Python 的特性

7.1.1 代码质量

Python 代码具有高度可读性，这使其更具可重用性和可维护性。它广泛支持高级软件工程规范，如面向对象和函数式编程。

7.1.2 提高开发人员工作效率

Python 具有干净优雅的编码风格。它使用类似英语的语法，并且是动态类型的。因此，你永远不需要声明一个变量，简单赋值即可将名称绑定到任何类型的对象。Python 代码明显少于等效的 C++或 Java 代码。这意味着更少的输入、调试和维护。与编译语言不同，Python 程序不需要编译和链接，这进一步提高了程序员的工作速度。

7.1.3 代码可移植性

由于 Python 是一种解释型语言，因而解释器必须管理可移植性任务。此外，Python 的解释器足够智能，可以在不同的平台上执行程序，以产生相同的输出。因此，你不需要更改任何一行代码。

7.1.4 内置和外部库

Python 打包了大量预构建和可移植的库集。可以在需要时加载它们来使用所需的功能。

7.1.5 组件集成

一些应用程序需要跨不同组件进行交互以支持端到端工作流。一个这样的组件可以是 Python 脚本，而其他组件是用 Java/C ++或任何其他技术编写的程序。

Python 有几种支持跨应用程序通信的方法。它允许加载 C 和 C++库，反之亦然，它具有与 Java 和.NET 组件集成、使用 COM/Silverlight 进行通信，以及通过串行端口与 USB 设备连接等机制。它甚至可以使用协议通过网络交换数据。

7.1.6 免费使用、修改和重新分发

Python 是一个 OSS。你可以免费使用它，也可以修改源代码并重新分发，甚至商用。正是由于这种开放性，Python 已经获得了一个庞大的社区基础，而且还在不断增长并增加价值。

7.1.7 面向对象的核心

Python 主要遵循面向对象程序设计。面向对象程序设计提供了一种构建代码的直观方式，对其背后概念的深入理解可以充分利用代码。

7.2 Python 编程使用领域

7.2.1 Web 应用程序开发

Python 在 Web 开发领域占有最大份额。许多雇主都在寻找了解 Python 的全栈程序员。所有这些都是使用 Python 开发的。通过学习 Django、Flask、CherryPy 和 Bottle 框架等，你可以成为其中的一员，这些框架为 Web 开发提供了广泛的支持。

7.2.2 科学和数字计算

Python 已经显著应用于科学和数字应用工作。这种进步有多种原因。首先，Python 是一种免费的开源语言，允许修改和重新分发其源代码。其次，Python 在科学和数字领域更具主导地位的原因是其专业模块数量的快速增长，而且所有这些都是免费提供的。

因此，Python 正在成为科学和数字领域的领导者。Python 语言的重点是提高生产力并提高可读性。

7.2.3 GUI 编程

Python 具有一些固有的特性，如干净简单的编码语法及动态类型支持。在开发复杂的 GUI 和图像处理应用程序时，它们可以作为催化剂。Python 干净的语法和众多的 GUI 库为程序员交付图形软件提供了巨大支持。

7.2.4 软件原型

Python 具有许多特性，使其成为原型设计的自然选择。一种特性是开源编程语言，有大量用户使用并为其开发做出贡献。此外，Python 重构代码使其具有轻量级、多功能、可伸缩和灵活等特点，这加速了初始原型的开发过程。而且，Python 提供了一个易于使用的界面来创建原型。

7.2.5 专业培训

Python 确实是适用于教学和培训的编程语言，可以成为初学者进入专业培训的垫脚石。初学者甚至可以学习数据分析和机器学习等领域的知识。因此，对能够教授基础和高级 Python 编程的专业培训师有着巨大的需求。你可以在课堂上离线培训或使用工具进行在线培训。

8. R 编程语言

在统计计算和图形领域，R 已成为非常流行的编程语言和开发环境。它起初是一个类似于 S 语言的 GNU 项目，但还有许多其他特性。它可用于计算各种统计测试，包括经典测试（如学生测试和相关性检验）。它拥有强大的用户群和强大的社区支持，所有人都可以免费使用。

8.1 什么是 R 编程语言？

R 编程语言由各种各样的统计和图形方法组成。它包含回归分析算法、机器学习、时间序列等。它的大多数包或库都是用 R 编写的。

为了提高效率，用 C、C++、FORTRAN、Python 和.NET 编写的程序也可以集成进来。R 语言在学术机构以及 Uber、Google 和 Facebook 等大公司中都很受欢迎。

R 语言的核心实际上是一种解释性编程语言，支持模块化编程、循环和分支。R 语言在数据分析中的使用通过以下步骤来实现：

- 编程。
- 转型。
- 发现。
- 建模。
- 传达结果。

R 语言编程环境基于命令行界面。

8.2 R 语言的特点

R 编程语言的不同特性如下：

- 简单有效，包含条件循环、递归函数和输入/输出工具。
- 其功能支持过程编程。通用函数支持面向对象程序设计。

- 支持矩阵算术运算。
- 作为解释语言，可以通过命令行操作。
- 提供存储和数据处理规范。
- 许多操作可用于处理数组、列表、矩阵和向量。
- 可以使用导出数据、处理数据库输入、数据查看、变量标签等功能。
- 它提供了显示图形分析结果的工具，可以在屏幕上显示，也可以作为硬拷贝输出。
- R 代码可以直接在控制台中运行，无须编译器。

Unit 5

Text A

Operating Systems

An operating system (OS) is system software that manages computer hardware and software resources and provides common services for computer programs. Application programs usually require an operating system to function.

1. Types of Operating Systems

1.1 Single-tasking and multi-tasking

A single-tasking operating system can only run one program at a time while a multi-tasking operating system allows more than one program to be running in concurrency.[1] This is achieved by time-sharing, where the available processor time is divided between multiple processes. These processes are each interrupted repeatedly in time slices by a task-scheduling subsystem of the operating system. Multi-tasking may be characterized in preemptive and cooperative types. In preemptive multitasking, the operating system slices the CPU time and dedicates a slot to each of the programs.

1.2 Single-user and multi-user

Single-user operating systems have no facilities to distinguish users, but may allow multiple programs to run in tandem. A multi-user operating system extends the basic concept of multi-tasking with facilities that identify processes and resources, such as disk space, belonging to multiple users, and the system permits multiple users to interact with the system at the same time.[2]

1.3 Distributed

A distributed operating system manages a group of distinct computers and makes them appear to be a single computer. The development of networked computers that could be linked and communicate with each other gave rise to distributed computing. Distributed computing are carried out on more than one machine. When computers in a group work in cooperation, they form a distributed system.

1.4 Embedded

Embedded operating systems are designed to be used in embedded computer systems. They are designed to operate on small machines like PDAs with less autonomy. They are able to operate with a limited number of resources. They are very compact and extremely efficient by design.

1.5 Real-time

A real-time operating system is an operating system that guarantees to process events or data by a specific moment in time. A real-time operating system may be single-tasking or multi-tasking, but when multi-tasking, it uses specialized scheduling algorithms so that a deterministic nature of behavior is achieved. An event-driven system switches between tasks based on their priorities or external events while time-sharing operating systems switch tasks based on clock interrupts.

2. Components

The components of an operating system all exist in order to make the different parts of a computer work together. User software needs to go through the operating system in order to use any of the hardware, whether it be as simple as a mouse or keyboard or as complex as an Internet component.[3]

2.1 Program execution

The operating system provides an interface between an application program and the computer hardware, so that an application program can interact with the hardware only by obeying rules and procedures programmed into the operating system. The operating system is also a set of services which simplify development and execution of application programs. Executing an application program involves the creation of a process by the operating system kernel which assigns memory space and other resources, establishes a priority for the process in multi-tasking systems, loads program binary code into memory, and initiates execution of the application program which then interacts with the user and with hardware devices.

2.2 Interrupts

Interrupts are central to operating systems, as they provide an efficient way for the operating system to interact with and react to its environment. Interrupt-based programming is directly supported by most modern CPUs. Interrupts provide a computer with a way of automatically saving local register contexts, and running specific code in response to events. Even very basic computers support hardware interrupts, and allow the programmer to specify code which may be run when that event takes place.

When an interrupt is received, the computer's hardware automatically suspends whatever program is currently running, saves its status, and runs computer code previously associated with the interrupt. This is analogous to placing a bookmark in a book in response to a phone call. In modern operating systems, interrupts are handled by the operating system's kernel. Interrupts may come from either the computer's hardware or the running program.

When a hardware device triggers an interrupt, the operating system's kernel decides how to deal with this event, generally by running some processing code. The amount of code being run depends on the priority of the interrupt. The processing of hardware interrupts is a task that is usually delegated to software called a device driver, which may be part of the operating system's kernel, part of another program, or both.[4] Device drivers may then relay information to a running program by various means.

A program may also trigger an interrupt to the operating system. If a program wishes to access hardware, for example, it may interrupt the operating system's kernel, which causes control to be passed back to the kernel. The kernel then processes the request. If a program wishes additional resources such as memory, it triggers an interrupt to get the kernel's attention.

2.3 Modes

Modern microprocessors (CPU or MPU) support multiple modes of operation. CPUs with this capability offer at least two modes: supervisor mode and user mode. In general terms, supervisor mode operation allows unrestricted access to all machine resources, including all MPU instructions. User mode operation sets limits on instruction use and typically disallows direct access to machine resources. CPUs might have other modes similar to user mode as well, such as the virtual modes in order to emulate older processor types, such as 16-bit processors on a 32-bit one, or 32-bit processors on a 64-bit one.

2.4 Memory management

Among other things, a multiprogramming operating system kernel must be responsible for managing all system memory which is currently in use by programs. This ensures that a program does not interfere with memory already in use by another program. Since programs share time, each program must have independent access to memory.

2.5 Virtual memory

The use of virtual memory addressing (such as paging or segmentation) means that the kernel can choose what memory each program may use at any given time, allowing the operating system to use the same memory locations for multiple tasks.

Virtual memory provides the programmer or the user with the perception that there is a much larger amount of RAM in the computer than is really there.

2.6 Multi-tasking

Multi-tasking refers to the running of multiple independent computer programs on the same computer, giving the appearance that it is performing the tasks at the same time. Since most computers can do at most one or two things at one time, this is generally done via time-sharing, which means that each program uses a share of the computer's time to execute.

2.7 Disk access and file systems

Access to data stored on disks is a central feature of all operating systems. Computers store data on disks using files, which are structured in specific ways in order to allow for faster access, higher reliability, and to make better use of the drive's available space. The specific way in which files are stored on a disk is called a file system, and it enables files to have names and attributes. It also allows them to be stored in a hierarchy of directories or folders arranged in a directory tree.

2.8 Device drivers

A device driver is a specific type of computer software developed to allow interaction with hardware devices. Typically this constitutes an interface for communicating with the device through the specific computer bus or communications subsystem that the hardware is connected to. It provides commands to and/or receives data from the device, and on the other end, the requisite interfaces to the operating system and software applications. It is a specialized hardware-dependent computer program which is also operating system specific that enables another program, typically an operating system or applications software package or computer program running under the operating system kernel, to interact transparently with a hardware device, and usually provides the requisite interrupt handling necessary for any necessary asynchronous time-dependent hardware interfacing needs.

New Words

time-sharing	['taɪm'ʃeərɪŋ]	*n.*分时
schedule	['ʃedju:l]	*n.*预定计划；时刻表，进度表
		*vt.*排定，安排
mass	[mæs]	*n.*大量，大多
		v.（使）集中，聚集
		adj. 大规模的；整个的；集中的
single-tasking	['sɪŋgl-'tɑ:skɪŋ]	*n.*单任务
concurrency	[kən'kʌrənsɪ]	*n.*并发（性）
achieve	[ə'tʃi:v]	*vt.*取得，获得；实现， 成功
		*vi.*达到预期的目的，实现预期的结果
multiple	['mʌltɪpl]	*adj.*多重的；多个的；多功能的
interrupt	[ˌɪntə'rʌpt]	*vt. & n.*中断
repeatedly	[rɪ'pi:tɪdlɪ]	*adv.*反复地，重复地
subsystem	[sʌb'sɪstəm]	*n.*子系统，分系统
preemptive	[prɪ'emptɪv]	*adj.*抢占式的
dedicate	['dedɪkeɪt]	*vt.*提供
single-user	['sɪŋgl-'ju:zə]	*n.*单用户
multi-user	['mʌltɪ-'ju:zə]	*n.*多用户
distinguish	[dɪ'stɪŋgwɪʃ]	*vi.*区分，辨别，分清
tandem	['tændəm]	*adv.*一个跟着一个地
multi-tasking	['mʌltɪ-'tɑ:skɪŋ]	*n.*多任务
permit	[pə'mɪt]	*v.*许可，准许
		*n.*许可，准许；许可证
distributed	[dɪs'trɪbju:tɪd]	*adj.*分布式的
distinct	[dɪ'stɪŋkt]	*adj.*有区别的
autonomy	[ɔ:'tɒnəmi]	*n.*自治
compact	[kəm'pækt]	*adj.*紧凑的；简洁的
guarantee	[ˌgærən'ti:]	*n.*保证，担保；保证人；抵押品
		*vt.*保证，担保
moment	['məʊmənt]	*n.*瞬间，片刻
algorithm	['ælgərɪðəm]	*n.*算法；运算法则
deterministic	[dɪˌtɜ:mɪ'nɪstɪk]	*adj.*确定性的
execution	[ˌeksɪ'kju:ʃn]	*n.*执行，实行，履行，执行

obey	[ə'beɪ]	*v.*服从，听从
development	[dɪ'veləpmənt]	*n.*开发，发展
assign	[ə'saɪn]	*vt.*分配，分派，选派
priority	[praɪ'ɒrətɪ]	*n.*优先，优先权
local	['ləʊkl]	*adj.*本地的
suspend	[sə'spend]	*v.*挂起，暂停
status	['steɪtəs]	*n.*状态
analogous	[ə'næləgəs]	*adj.*相似的，可比拟的
bookmark	['bʊkmɑ:k]	*n.*书签 *vt.*给……设置书签
trigger	['trɪgə]	*vt.*引发，触发
relay	['ri:leɪ]	*vt.*转发，转播，传达
attention	[ə'tenʃn]	*n.*注意，注意力
supervisor	['su:pəvaɪzə]	*n.*管理者；监督者；指导者
unrestricted	[ˌʌnrɪ'strɪktɪd]	*adj.*不受限制的
emulate	['emjuleɪt]	*vt.*仿真
multiprogramming	['mʌltɪˌprəʊgræmɪŋ]	*n.*多道程序设计，多程序设计
paging	['peɪdʒɪŋ]	*n.*分页
segmentation	[ˌsegmen'teɪʃn]	*n.*分割；分段；分节
perception	[pə'sepʃn]	*n.*知觉，觉察
disk	[dɪsk]	*n.*磁盘
reliability	[rɪˌlaɪə'bɪlətɪ]	*n.*可靠性
hierarchy	['haɪərɑ:kɪ]	*n.*分层，层次
directory	[də'rektərɪ]	*n.*目录
constitute	['kɒnstɪtju:t]	*vt.*构成，组成；制定，设立
transparently	[træns'pærəntlɪ]	*adv.*显然地，易觉察地
asynchronous	[eɪ'sɪŋkrənəs]	*adj.*异步的
time-dependent	[taɪm-dɪ'pendənt]	*n.*时间相关；时间依赖性 *adj.*时间依赖的
hardware-dependent	['hɑ:dweə-dɪ'pendənt]	*n.*硬件相关；硬件依赖性 *adj.*硬件依赖的

Phrases

time-sharing operating system	分时操作系统
cost allocation	成本分摊，成本分配
mass storage	大容量存储器
time slice	时间片
be characterized in ...	以……为特征
single-user operating system	单用户操作系统

multi-user operating system	多用户操作系统
disk space	磁盘空间
interact with ...	与……相互作用，与……交互; 与……相互配合
at the same time	同时；一起
distributed operating system	分布式操作系统
embedded operating system	嵌入式操作系统
real-time operating system	实时操作系统
event-driven system	事件驱动系统
clock interrupt	时钟中断
work together	合作，协作，一起工作
memory space	存储空间
device driver	设备驱动程序
user mode	用户态
supervisor mode	管态
virtual mode	虚拟模式
multiprogramming operating system	多程序操作系统
multiple task	多任务
file system	文件系统
directory tree	目录树
applications software package	应用软件包

Abbreviations

PDA (Personal Digital Assistant)	个人数字助理，掌上电脑
MPU (Micro Processor Unit)	微处理器单元

Notes

[1] A single-tasking system can only run one program at a time while a multi-tasking operating system allows more than one program to be running in concurrency.

本句中，while 连接两个并列句子，表示“对比”关系，意思是“而，然而”。

例如：Some programmers like C while others prefer Python.

[2] A multi-user operating system extends the basic concept of multi-tasking with facilities that identify processes and resources, such as disk space, belonging to multiple users, and the system permits multiple users to interact with the system at the same time.

本句中，that identify processes and resources, such as disk space, belonging to multiple users 是一个定语从句，修饰和限定 facilities；在该从句中，belonging to multiple users 是一个现在分词短语，作定语，修饰和限定 resources。

[3] User software needs to go through the operating system in order to use any of the hardware, whether it be as simple as a mouse or keyboard or as complex as an Internet component.

本句中，in order to use any of the hardware 是目的状语。whether it be as simple as a mouse or keyboard or as complex as an Internet component 是一个让步状语从句，修饰主句的谓语 needs。

[4] The processing of hardware interrupts is a task that is usually delegated to software called a device driver, which may be part of the operating system's kernel, part of another program, or both.

本句中，that is usually delegated to software called a device driver, which may be part of the operating system's kernel, part of another program, or both 是一个定语从句，修饰和限定 a task；在该从句中，called a device driver 是一个过去分词短语，作定语，修饰和限定 software。which may be part of the operating system's kernel, part of another program, or both 是一个非限定性定语从句，对 device driver 进行补充说明。

Text B

The Elements of a Database

What elements comprise a database? This article deals mainly with the objects that comprise a database. Several concepts are worthy of coverage within the scope of the database as it relates to database design. As you work with data and databases, you will see how the origination of business information and databases is formulated into database elements. The intent here is to provide a brief coverage of basic database elements to provide you with a basic understanding of the elements found in a database.

1. Table

A table is the primary unit of physical storage for data in a database. When a user accesses the database, a table is usually referenced for the desired data. Multiple tables might comprise a database, therefore a relationship might exist between tables. Because tables store data, a table requires physical storage on the host computer for the database.

Figure 5-1 illustrates tables in a schema. Each table in the figure is related to at least one other table. Some tables are related to multiple tables.

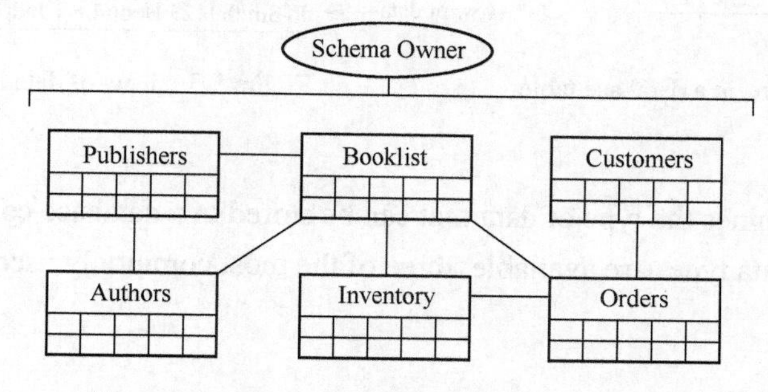

Figure 5-1 Database tables and their relationships

There are four types of tables which are commonly used:

- Data tables store most of the data found in a database.

- Join tables are tables used to create a relationship between two tables that would otherwise be unrelated.
- Subset tables contain a subset of data from a data table.
- Validation tables, often referred to as code tables, are used to validate data entered into other database tables.

Tables are used to store the data that the user needs to access. Tables might also have constraints attached to them, which control the data allowed to be entered into the table. An entity from the business model is eventually converted into a database table.

2. Columns

A column, or field, is a specific category of information that exists in a table. A column is to a table what an attribute is to an entity. In other words, when a business model is converted into a database model, entities become tables and attributes become columns. A column represents one related part of a table and is the smallest logical structure of storage in a database. Each column in a table is assigned a data type. The assigned data type determines what type of values that can populate a column. When visualizing a table, a column is a vertical structure in the table that contains values for every row of data associated with a particular column.

In Figure 5-2, columns within the Customers table are shown. Each column is a specific category of information. All of the data in a table associated with a field is called a column.

3. Rows

A row of data is the collection of all the columns in a table associated with a single occurrence. Simply speaking, a row of data is a single record in a table. For example, if there are 25,000 book titles with which a bookstore deals, there will be 25,000 records, or rows of data, in the book titles table once the table is populated. The number of rows within the table will obviously change as books' titles are added and removed. See Figure 5-3 for an illustration of a row of data in a table.

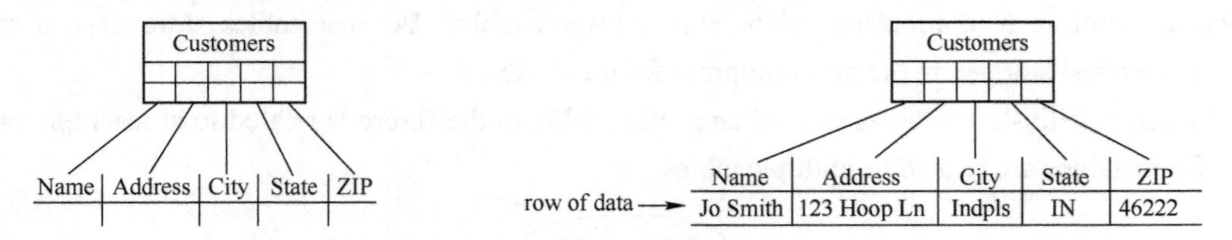

Figure 5-2 Columns in a database table

Figure 5-3 Row of data in a database table

4. Data Types

A data type determines the type of data that can be stored in a database column.

Although many data types are available, three of the most commonly used data types are

- Alphanumeric.
- Numeric.
- Date and time.

Alphanumeric data types are used to store characters, numbers, special characters, or nearly any combination. If a numeric value is stored in an alphanumeric field, the value is treated as a character, not a number. In other words, you should not attempt to perform arithmetic functions on numeric values stored in alphanumeric fields. Numeric data types are used to store only numeric values. Date and time data types are used to store date and time values, which widely vary depending on the relational database management system (RDBMS) being used.

5. Keys

The integrity of the information stored in a database is controlled by keys. A key is a column value in a table that is used to either uniquely identify a row of data in a table, or establish a relationship with another table. A key is normally correlated with one column in table, although it might be associated with multiple columns. There are two types of keys: primary keys and foreign keys.

5.1 Primary keys

A primary key is the combination of one or more column values in a table that make a row of data unique within the table. Primary keys are typically used to join related tables. Even if a table has no child table, a primary key can be used to disallow the entry of duplicate records into a table.

5.2 Foreign keys

A foreign key is the combination of one or more column values in a table that reference a primary key in another table. Foreign keys are defined in child tables. A foreign key ensures that a parent record has been created before a child record. Conversely, a foreign key also ensures that the child record is deleted before the parent record.

6. Relationships

Most databases are divided into many tables, most of which are related to one another. In most modern databases, such as the relational database, relationships are established through the use of primary and foreign keys. The purpose of separating data into tables and establishing table relationships is to reduce data redundancy. The process of reducing data redundancy in a relational database is called normalization.

Three types of table relationships that can be derived are as follows:

- One-to-one. One record in a table is related to only one record in another table.
- One-to-many. One record in a table can be related to many records in another table.
- Many-to-many. One record in a table can be related to one or more records in another table, and one or more records in the second table can be related to one or more records in the first table.

Figure 5-4 briefly illustrates table relationships in a relational database. A relational database allows parent tables to have many child tables, and child tables to have many parent tables. The figure shows two tables. Table 1 has an ID column (primary key) and Table 2 has an FK column (foreign key). In the one-to-one relationship example, notice that for every ID in Table 1, there is only one ID in Table 2. In the one-to-many relationships example, notice that the ID of Table 1 has many occurrences in Table 2. In the many-to-many relationship example, notice that the ID in Table 1 might occur multiple times in Table 2 as a foreign key, and the ID in Table 2 might occur multiple times in Table 1.

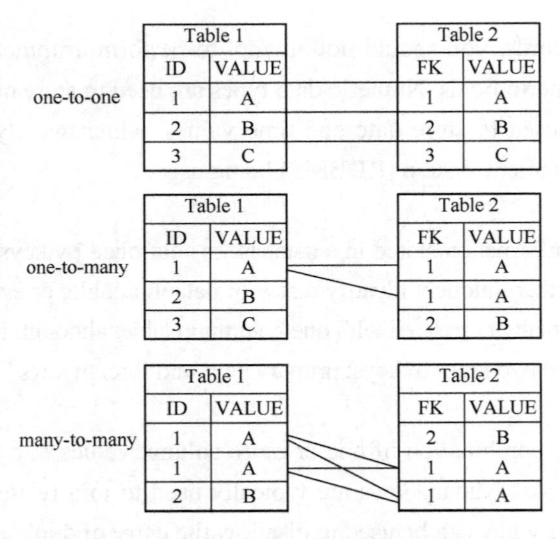

Figure 5-4 Available table relationships in the relational database model

7. Conclusion

The intent of this article was to provide a brief coverage of basic database elements to provide you with a basic understanding of the elements found in a database. While it may not seem like a lot has been covered, you have actually now been exposed to the key topics involved in working with databases. As you continue to work with databases, you will find that you will continually be confronted with the above terms and concepts, including schemas, tables, columns, rows, and keys.

New Words

database ['deɪtəbeɪs] *n.*数据库

comprise [kəm'praɪz] *v.*包含，由……组成

concept ['kɒnsept] *n.*概念，观念

formulate ['fɔ:mjʊleɪt] *vt.*规划，设计

table ['teɪbl] *n.*表，表格
*vt.*制表

relationship [rɪ'leɪʃnʃɪp] *n.*关联，关系

schema ['ski:mə] *n.*模式，大纲，模型

join [dʒɔɪn] *vt.*连接，结合，参加，加入

subset ['sʌbset] *n.*子集

validation [ˌvælɪ'deɪʃn] *n.*确认，有效性

constraint [kən'streɪnt] *n.*强制，约束

column ['kɒləm] *n.*列，栏

field [fi:ld] *n.*域；字段

attribute [æ'trɪbju:t] *n.*属性，品质，特征

row [rəʊ] *n.*行，排

collection	[kə'lekʃn]	*n.*收集，收取
alphanumeric	[ˌælfənju: 'merɪk]	*adj.*字符数字的，字母数字混合编制的
numeric	[nju:'merɪk]	*adj.*数字的
character	['kærəktə]	*n.*字符
arithmetic	[ə'rɪθmətɪk]	*n.*算术，算法
uniquely	[jʊ'ni:klɪ]	*adv.*独特地，唯一地
identify	[aɪ'dentɪfaɪ]	*vt.*标识，识别，鉴别
establish	[ɪ'stæblɪʃ]	*vt.*建立，设立，安置，确定
disallow	[ˌdɪsə'laʊ]	*vt.*不准许，禁止，不接受
duplicate	['dju:plɪkeɪt]	*adj.*复制的，两重的，两倍的，完全相同 *n.*复制品，副本 *vt.*复写，复制，使加倍，使成双
reference	['refrəns]	*n.*提及，涉及，参考
record	['rekɔ:d]	*n.*记录
separating	['sepəreɪtɪŋ]	*adj.*分开的，分离的
redundancy	[rɪ'dʌndənsi]	*n.*冗余
normalization	[ˌnɔ:məlaɪ'zeɪʃn]	*n.*归一化，正规化，标准化，规格化，规范化
illustrate	['ɪləstreɪt]	*vt.*举例说明，图解，加插图于，阐明

Phrases

physical storage	物理存储，实际存储
host computer	主机
database model	数据库模型
logical structure	逻辑结构
associate with …	同……联合
special character	特殊字符
relational database management system (RDBMS)	关系数据库管理系统
child table	子表

Word Building

在一个单词的前面或后面加上一定的词缀来构成一个新词的方法被称为“派生法”。词缀有前缀与后缀两种。

前缀由一个或几个字母组成，放在词根或单词之前，与词根或单词构成一个新词。每一个前缀都有一定的含义。前缀一般不超过 5～6 个字母。加了前缀的单词其词性一般不发生变化，只改变原来的意思。常用的前缀如下：

1. a-表示意义无、非、不

periodic 周期的 —— aperiodic 非周期的

centric 中心的 —— acentric 无中心的

2. ab-表示：离去、脱离
normal 正常的 —— abnormal 不正常的
3. anti-表示：反对、相反、防止、防治
virus 病毒 —— anti-virus 防病毒
magnetic 磁性的 —— antimagnetic 防磁的
war 战争 —— antiwar 反战的
4. auto-表示：自动、自身
alarm 报警器 —— autoalarm 自动报警器
rotation 转动、旋转 —— autorotation 自动旋转,自动
5. bi-表示：两、二、重
coloured 颜色的 —— bicoloured 双色的
fold 倍 —— bifold 两倍的
monthly 每月的 —— bimonthly 两月一次的,双月刊
6. by-表示：边、侧、偏、副、非正式
product 产品 —— by-product 副产品
effect 作用 —— by-effect 副作用
road 路 —— byroad 小路
7. centi-表示：百分之一
metre 米 —— centimetre 厘米
grade 度 —— centigrade 百分度
8. co-表示：共同、相互
operation 操作 —— co-operation 合作
run 管理 —— co-run 共同管理

Career Training

名　片

名片，又叫商业卡或访问卡，一般包括某人的工作单位、职业头衔、邮政编码、电话号码等个人信息。

请看以下示例：

Del China Ltd. Shanghai Branch

Wang Jianguo
Sales Manager
Sales Department 28th floor Happy Valley Complex
105 Bao Shan Road Shanghai,China　Post Code: 200003
Tel: 021-5376888 Fax: 021-5376358　E-mail: jianguo @ hotmail.com

德尔中国股份有限公司上海分公司

王建国
销售部经理
地址：上海宝山路 105 号幸福谷综合楼 28 楼销售部
邮编：200003　传真：0755-5376358
电话：021-5376888　E-mail：jianguo @ hotmail.com

中文名片如下：

南京光明电子公司

张卫宏

注册电子工程师

地址：南京珠江路友谊大厦 518 室　邮编：　210001

电话：025-65618888　　E-mail：weihong @ yeah.net

其英文名片如下：

Guang Ming Electrical Company

Zhang Weihong

Certified Electrical Engineer

Room 518 Friendship Building　Zhujiang Road, Nanjing, P.R.China

Post Code: 210001　Tel:025-65618888

E-mail: weihong @ yeah.net

软件水平考试试题解析

【真题再现】

从供选择的答案中选出应填入下列英语文句中____内的正确答案，把编号写在答卷的对应栏内。

The UNIX system contains several __A__ that comply with the definition of a software tool. Among them are programs that __B__ and manipulate text, programs that analyze text files, and programs that format text files to produce high quality hard copy suitable for __C__.

One characteristic of these tools is that they operate on ordinary text __D__, which means that you can read the input and output files by simply listing them on a __E__.

供选择的答案：

A～E：①terminal　②keyboard　③programs　④programming　⑤files

⑥directories　⑦create　⑧build　⑨publication　⑩painting

【答案】 A：③　B：⑦　C：⑨　D：⑤　E：①

【试题解析】

A：本句的意思是“UNIX 系统包含几个符合软件工具定义的程序”。根据第二个句子可以推出此处的中文含义应是“程序”。故选③。

B：根据句子结构，此处需要填写一个动词，因此只能在 create 和 build 之间进行选择，建立一个文本用 create，而不能用 build。故选⑦。

C：suitable for publication 的意思是“适合出版”。根据句意，选⑨。

D：text files 的意思是“文本文件”。根据句意，选⑤。

E：本句的意思是“通过在终端上列表，即可阅读输入和输出文件”，此处的中文含义应是“终端”。故选①。

【参考译文】

UNIX 系统包含几个符合软件工具定义的程序。其中有创建和管理文本的程序、分析文本文件的程序、产生适合出版的高质量硬拷贝的格式化文本文件的程序。这些工具都有一个特点，就是它们对普通的文本文件进行操作，这意味着在终端上列出输入和输出文件即可阅读它们。

Exercises

[Ex. 1] 根据 Text A 回答以下问题。

1) What is an operating system?

2) What is the difference between a single-tasking system and a multi-tasking operating system?

3) What does a distributed operating system do?

4) What are embedded operating systems designed to do?

5) What is a real-time operating system?

6) What do interrupts do?

7) What does the computer's hardware do when an interrupt is received?

8) What does supervisor mode operation do in general terms?

9) What does the use of virtual memory addressing (such as paging or segmentation) mean?

10) What is a device driver?

[Ex. 2] 根据 Text B 填空。

1) A table is the primary unit of ________________ in a database. When a user accesses the database, a table is usually referenced for ________________.

2) The four types of tables which are commonly used are ____________, ____________, ____________ and ____________.

3) A column represents one related part of ____________ and is ____________ of storage in a ___________.

4) A row of data is the collection of ____________ in a table associated with a single occurrence.

5) Three of the most commonly used data types are ___________, ____________ and ____________.

6) Alphanumeric data types are used to store ___________, ___________, ___________, or___________ .

7) The integrity of the information stored in a database is controlled by __________. A key is ____________ in a table that is used to either uniquely identify ____________ in a table, or

____________ with another table.

8) A primary key is the combination of ____________ in a table that make ____________ within the table.

9) In most modern databases, such as the relational database, relationships are established through the use of ______________. The purpose of separating data into tables and establishing table relationships is to ______________. The process of reducing data redundancy in a relational database is called______________.

10) Three types of table relationships that can be derived are ____________, ____________ and ____________.

[Ex. 3] 把下列句子翻译为中文。

1) A computer with the ability to execute several programs simultaneously is called a time-sharing system.

2) The more primitive single-tasking operating systems can run only one process at a time.

3) The primary concerns in this kind of application are concurrency and atomicity.

4) This is not what you want, because that would interrupt the application's normal workflow.

5) If you use Approach 1, you develop one subsystem at a time.

6) How can you restrict access to files in a multi-user environment?

7) Each process in a multi-tasking OS runs in its own memory sandbox.

8) Because of the distributed firewall distribution across the entire enterprise network or the server, so it has unlimited extension ability.

9) What's the complexity of this algorithm?

10) We perform an analogous process on the server-side.

[Ex. 4] 软件水平考试真题自测。

从供选择的答案中选出应填入下列英语文句中____内的正确答案，把编号写在答卷的对应栏内。

The development of the Semantic Web proceeds in steps, each step building a layer on top of another. The pragmatic justification for this approach is that it is easier to achieve ___1___ on small steps, whereas it is much harder to get everyone on board if too much is attempted. Usually there are several research groups moving in different directions; this ___2___ of ideas is a major driving force for scientific progress. However, from an engineering perspective there is a need to standardize. So, if most researchers agree on certain issues and disagree on others, it makes sense to fix the point of agreement. This way, even if the more ambitious research efforts should fail, there will be at least ____3____ positive outcomes. Once a ____4____ has been established, many more groups and companies will adopt; it, instead of waiting to see which of the alternative research lines will be successful in the end. The nature of the Semantic Web is such that companies and single users must build tools, add content, and use that content. We cannot wait until the full Semantic Web vision materialize. It may take another ten years for it to be realized to its full ____5____ (as envisioned today, of course).

(1) A. conflicts B. consensus C. success D. disagreement
(2) A. competition B. agreement C. cooperation D. collaboration
(3) A. total B. complete C. partial D. entire
(4) A. technology B. standard C. pattern D. model
(5) A. area B. goal C. object D. extent

[Ex. 5] 听短文填空。

Exercises 5

A Network Operating System (NOS) is a software program that controls other software and hardware running on a ____1____. It also allows multiple computers, known as network computers, to communicate with one central hub and share ____2____, run applications, and send messages. Such a system can consist of a wireless network, Local Area Network (LAN), or even two or three computer networks ____3____ together. Administrators running these networks typically have training in different network operating ____4____.

Networks usually consist of multiple computers connected to each other through a central hub or router. This central hub, in turn, may be connected to a ____5____ main computer. Networks can also include other ____6____ like printers, data backup systems, and central storage facilities. The main network computer monitors all the connected machines with the help of the network operating system ____7____.

A network operating system often has a menu-based administration ____8____. From this interface, a network administrator can perform a variety of activities. He or she uses the interface to ____9____ hard drives, set up security permissions, and establish log-in information for each user. An administrator can also use the interface of a network operating system to ____10____ shared printers and configure the system to automatically back up data on a scheduled basis.

Reference Translation

操 作 系 统

操作系统（OS）是管理计算机硬件和软件资源，并为计算机程序提供公共服务的系统软件。应用程序通常需要一个操作系统才能运行。

1. 操作系统的类型

1.1 单任务和多任务

单任务操作系统一次只能运行一个程序；而多任务操作系统允许多个程序以并发方式运行，这是通过分时实现的，即把可用的处理器时间分配给多个进程。这些进程都由操作系统的任务调度子系统在时间片中重复中断。多任务处理具有抢占式和协作式特点。在抢占式多任务处理中，操作系统会把 CPU 时间“切片”，并为每个程序分配一个时隙。

1.2 单用户和多用户

单用户操作系统没有区分用户的功能，但可以允许多个程序串联运行。多用户操作系统扩展了多任务的基本概念，其中包括识别属于多个用户的进程和资源（例如磁盘空间）的设备，并且允许多个用户同时与操作系统交互。

1.3 分布式

分布式操作系统管理一组不同的计算机，使它们看起来像是一台计算机。可以相互联结和通信的联网计算机的开发促进了分布式计算的产生。分布式计算在多台计算机上执行。当一个组中的计算机协同工作时，它们形成一个分布式系统。

1.4 嵌入式

嵌入式操作系统设计用于嵌入式计算机系统。它们被设计成运行在 PDA 等小型机器上，自主性较低；能够使用有限数量的资源进行操作；结构非常紧凑，设计得非常高效。

1.5 实时

实时操作系统是可以保证在特定时刻及时处理事件或数据的操作系统。实时操作系统可以是单任务也可以是多任务，但当进行多任务处理时，它使用专门的调度算法以实现行为的确定性。事件驱动系统根据优先级或外部事件切换任务，而分时操作系统根据时钟中断切换任务。

2. 组件

为使计算机各个部分协同工作，操作系统有多种组件。不管是简单地使用鼠标或键盘，还是复杂地使用因特网等组件，所有用户软件要使用任何硬件都要通过操作系统来实现。

2.1 程序执行

操作系统提供应用程序和计算机硬件之间的接口，这样应用程序只要遵守操作系统的编程规则和过程就可以与硬件交互。操作系统也是一组简化应用程序开发和执行的服务。

应用程序的执行涉及：操作系统内核创建进程，为该进程分配内存空间和其他资源，在多任务操作系统中建立进程的优先级，将程序二进制代码加载到内存中，启动应用程序的执行。应用程序被执行后，就能与用户和硬件设备进行交互了。

2.2 中断

中断是操作系统的核心，因为它们为操作系统提供了一种与环境交互并对环境做出反应的有效方式。大多数现代 CPU 都直接支持基于中断的编程。中断为计算机提供了一种自动保存本地寄存器状态，并运行特定代码以响应事件的方法。即使是非常基本的计算机，也支持硬件中断，并允许程序员指定在该事件发生时可以运行的代码。

当收到中断时，计算机的硬件会自动暂停当前正在运行的程序，保存其状态，并运行先前与中断相关的计算机代码；这类似于在书中放置书签以响应电话呼叫。在现代操作系统中，中断由操作系统的内核来处理。中断可能来自计算机的硬件，也可能来自正在运行的程序。

当硬件设备触发中断时，操作系统的内核决定如何处理此事件，通常会运行一些处理代码。处理代码的代码量取决于中断的优先级。硬件中断通常由被称为设备驱动程序的软件来处理；该软件可能是操作系统内核的一部分，也可能是另一个程序的一部分，或两者兼而有之。设备驱动程序可以通过各种方式将信息转发到正在运行的程序。

程序还可以触发操作系统的中断。例如，如果程序希望访问硬件，它可能会中断操作系统的内核，从而使内核得到控制权，以便内核处理程序的请求。如果程序需要额外的资源（如内存），它会触发中断以引起内核的注意。

2.3　模式

现代微处理器（CPU 或 MPU）支持多种操作模式。支持多种操作模式的 CPU 提供至少两种模式：管态和用户态。一般而言，管态允许不受限制地访问所有机器资源，包括所有 MPU 指令。用户态设置了对指令使用的限制，通常不允许直接访问机器资源。CPU 也可能具有与用户态类似的其他模式，例如，虚拟模式以模拟较旧的处理器类型，如在 32 位处理器上模拟 16 位处理器或在 64 位处理器上模拟 32 位处理器。

2.4　内存管理

除此之外，多程序设计操作系统内核必须负责管理程序当前使用的所有系统内存。这可以确保程序不会使用已被另一个程序使用的内存。由于程序共享时间，每个程序必须具有独立的内存访问权限。

2.5　虚拟内存

使用虚拟内存寻址（例如分页或分段）意味着内核可以指定每个程序在任何给定时间可以使用的内存，从而允许操作系统将相同的内存位置用于多个任务。

虚拟内存让程序员或用户感觉到，计算机中的 RAM 数量远远超过实际存在的数量。

2.6　多任务处理

多任务是指在同一台计算机上运行多个独立的计算机程序，看起来就像同时执行这些任务。由于大多数计算机一次最多只能完成一两件事，因而多任务处理通常是通过分时来完成的，这意味着每个程序都使用计算机执行时间的一部分。

2.7　磁盘访问和文件系统

访问存储在磁盘上的数据是所有操作系统的核心功能。计算机以文件的方式将数据存储在磁盘上，这些文件以特定方式构建，以便提高访问速度和可靠性，并更好地利用驱动器的可用空间。文件存储在磁盘上的具体方法被称为文件系统，它使文件具有名称和属性；还允许把文件存储在层次结构的目录树中，或者存储在目录树的文件夹中。

2.8　设备驱动程序

设备驱动程序是用来与硬件设备交互的特定类型的计算机软件。通常，它构成了用于与设备通信的接口。通过与硬件连接的特定计算机总线或通信子系统，设备驱动程序向设备发布命令和/或从设备接收数据，另外也为操作系统和应用程序提供必要的接口。它是一种依赖于硬件的专用计算机程序，也与操作系统相关联，使另一个程序（通常是操作系统或应用程序软件包或在操作系统内核下运行的计算机程序）能够与硬件设备透明地交互，并且通常为任何异步依赖时间的硬件接口需求提供必要的中断处理。

Unit 6

Text A

Networking Devices (1)

1. Hub

Networks using a star topology require a central point for the devices to connect. Originally this device was called a concentrator since it consolidated the cable from all network devices. The basic form of concentrator is the hub.

Text A Networking Devices (1)

The hub is a hardware device that contains multiple, independent ports that match the cable type of the network.[1] Most common hubs interconnect Category 3 or 5 twisted pair cable with RJ45 ends, although Coax BNC and Fiber Optic BNC hubs also exist. Hubs offer an inexpensive option for transporting data between devices, but hubs don't offer any form of intelligence.

Hubs can be active or passive. An active hub strengthens and regenerates the incoming signals before sending the data to its destination. Passive hubs do nothing with the signal.

1.1 Ethernet Hubs

An Ethernet hub is also called a multiport repeater. A repeater is a device that amplifies a signal as it passes through it, to counteract the effects of attenuation. If, for example, you have a thin Ethernet network with a cable segment longer than the prescribed maximum of 185 meters, you can install a repeater at some point in the segment to strengthen the signals and increase the maximum segment length.[2] This type of repeater only has two BNC connectors, and is rarely seen these days.

The hubs used on UTP Ethernet networks are repeaters as well, but they can have many RJ45 ports instead of just two BNC connectors. When data enters the hub through any of its ports, the hub amplifies the signal and transmits it out through all of the other ports. This enables a star network to have a shared medium, even though each computer has its own separate cable. The hub relays every packet transmitted by any computer on the network to all of the other computers and it also amplifies the signals.

The maximum segment length for a UTP cable on an Ethernet network is 100 meters. A segment is defined as the distance between two communicating computers. However, because the hub also functions as a repeater, each of the cables connecting a computer to a hub port can be up to 100 meters long, allowing a segment length of up to 200 meters when one hub is inserted in the network.

1.2 Multi-station Access Unit

A Multi-station Access Unit (MAU) is a special type of hub used for token ring networks. The word "hub" is most often used in relation to Ethernet networks, and MAU only refers to token ring networks. On the outside, the MAU looks like a hub. It connects to multiple network devices, each with a separate cable.

Unlike a hub that uses a logical bus topology over a physical star, the MAU uses a logical ring topology over a physical star.

When the MAU detects a problem with a connection, the ring will beacon. Because it uses a physical star topology, the MAU can easily detect which port the problem exists on and close the port, or "wrap" it. The MAU does actively regenerate signals as it transmits data around the ring.[3]

2. Switches

Switches are a special type of hub that offers an additional layer of intelligence to basic, physical-layer repeater hubs. A switch must be able to read the MAC address of each frame it receives. This information allows switches to repeat incoming data frames only to the computer or computers to which a frame is addressed. This speeds up the network and reduces congestion(Figure 6-1).

Figure 6-1 Switch

Switches operate at both the physical layer and the data link layer of the (OSI) model.

3. Bridges

A bridge is used to join two network segments together, it allows computers on either segment to access resources on the other. They can also be used to divide large networks into smaller segments. Bridges have all the features of repeaters, but can have more nodes, and since the network is divided, there is fewer computers competing for resources on each segment thus improving network performance.

Bridges can also connect networks that run at different speeds, different topologies, or different protocols. But they cannot join an Ethernet segment with a token ring segment because these use different networking standards. Bridges operate at both the physical layer and the MAC sublayer of the data link layer. Bridges read the MAC header of each frame to determine on which side of the bridge the destination device is located, the bridge then repeats the transmission to the segment where the device is located.

4. Routers

Routers are networking devices used to extend or segment networks by forwarding packets from one logical network to another. Routers are most often used in large internetworks that use the TCP/IP protocol suite and for connecting TCP/IP hosts and local area networks (LANs) to the Internet using dedicated leased lines.

Routers work at the network layer (layer 3) of the OSI model for networking to move packets between networks using their logical addresses (which, in the case of TCP/IP, are the IP addresses of destination hosts on the network). Because routers operate at a higher OSI level than bridges do, they have better packet-routing and filtering capabilities and greater processing power, which results in routers costing more than bridges.

Routers contain internal tables of information called routing tables that keep track of all known network addresses and possible paths throughout the internetwork, along with cost of reaching each network.[4] Routers route packets based on the available paths and their costs, thus taking advantage of redundant paths that can exist in a mesh topology network.

Because routers use destination network addresses of packets, they work only if the configured network protocol is a routable protocol such as TCP/IP or IPX/SPX. This is different from bridges, which are protocol independent. The routing tables are the heart of a router; without them, there's no way for the router to know where to send the packets it receives.[5]

Unlike bridges and switches, routers cannot compile routing tables from the information in the data packets they process. This is because the routing table contains more detailed information than that is found in a data packet, and also because the router needs the information in the table to process the first packets it receives after being activated. A router can't forward a packet to all possible destinations in the way that a bridge can.

Static routers: These must have their routing tables configured manually with all network addresses and paths in the internet work.

Dynamic routers: These automatically create their routing tables by listening to network traffic.

Routing tables are the means by which a router selects the fastest or nearest path to the next "hop" on the way to a data packet's final destination. This process is done through the use of routing metrics.

Routing metrics are the means of determining how much distance or time a packet will require to reach the final destination. Routing metrics are provided in different forms.

You can use routers to segment a large network and to connect local area segments to a single network backbone that uses a different physical layer and data link layer standard. They can also be used to connect LAN's to a WAN's.

5. Brouters

Brouters are a combination of router and bridge. This is a special type of equipment used for networks that can be either bridged or routed, based on the protocols being forwarded. Brouters are complex, fairly expensive pieces of equipment and as such are rarely used.

A brouter transmits two types of traffic at the exact same time: bridged traffic and routed traffic. For bridged traffic, the brouter handles the traffic the same way a bridge or switch would, forwarding data based on the physical address of the packet. This makes the bridged traffic fairly fast, but slower than if it were sent directly through a bridge because the brouter has to determine whether the data packet should be bridged or routed.

6. Gateways

A gateway is a device used to connect networks using different protocols. Gateways operate at the network layer of the OSI model. In order to communicate with a host on another network, an IP host must be configured with a route to the destination network. If a configuration route is not found, the host uses the gateway (default IP router) to transmit the traffic to the destination host. The default gateway is where the IP sends packets that are destined for remote networks. If no default gateway is specified, communication is limited to the local network. Gateways receive data from a network using one type of protocol stack, remove that protocol stack and repackage it with the protocol stack that the other network can use.

Examples:

- E-mail gateways, a gateway that receives simple mail transfer protocol (SMTP) e-mail, translates it into a standard X.400 format, and forwards it to its destination.
- Gateway service for netware (GSNW), which enables a machine running Microsoft Windows NT server or Windows server to be a gateway for Windows clients so that they can access file and print resources on a NetWare server
- Gateways between a systems network architecture (SNA) host and computers on a TCP/IP network, such as the one provided by Microsoft SNA server.
- A packet assembler/disassembler (PAD) that provides connectivity between a local area network (LAN) and an X.25 packet-switching network.

New Words

connect	[kə'nekt]	*v*.连接，联合
concentrator	['kɒnsentreɪtə]	*n*.集中器
consolidate	[kən'sɒlɪdeɪt]	*vt*.把……合成一体，合并 *vi*.合并，联合
independent	[ˌɪndɪ'pendənt]	*adj*.独立的，不受约束的
inexpensive	[ˌɪnɪk'spensɪv]	*adj*.便宜的，不贵重的
transport	['trænspɔːt]	*n*.传送器，运输 *vt*.传送，运输
active	['æktɪv]	*adj*.主动的，活动的
passive	['pæsɪv]	*adj*.被动的
strengthen	['streŋθn]	*v*.加强，巩固

regenerate	[rɪ'dʒenəreɪt]	*vt.*使新生，重建
		*vi.*新生，再生
		*adj.*新生的，更新的
send	[send]	*vt.*送，寄，发送
multiport	['mʌltɪpɔ:t]	*adj.*多端口的
repeater	[rɪ'pi:tə]	*n.*中继器
amplify	['æmplɪfaɪ]	*vt.*放大，增强
counteract	[ˌkaʊntər'ækt]	*vt.*抵消，阻碍
attenuation	[əˌtenjʊ'eɪʃn]	*n.*衰减
segment	['segmənt]	*n.*段，节，片断
		*v.*分割
prescribe	[prɪ'skraɪb]	*v.*指示，规定
separate	['sepərət]	*adj.*分开的，分离的，单独的
	['sepəreɪt]	*v.*分开，隔离，分散
multistation	[mʌltɪs'teɪʃn]	*n.*多站
detect	[dɪ'tekt]	*vt.*察觉，发现，探测
wrap	[ræp]	*vt.*遮蔽，隐藏
switch	[swɪtʃ]	*n.*交换机，开关
frame	[freɪm]	*n.*帧，画面，框架
congestion	[kən'dʒestʃən]	*n.*拥塞
protocol	['prəʊtəkɒl]	*n.*协议
sublayer	['sʌb'leɪə]	*n.*子层，下层
router	['ru:tə]	*n.*路由器
internetwork	[ɪntə:r'netwɜ:k]	*n.*网间
filter	['fɪltə]	*n.*滤波器，滤波，筛选
		*vt.*过滤
reach	[ri:tʃ]	*vt.*到达，达到，伸出，延伸
forward	['fɔ:wəd]	*vt.*转发，转寄
manually	['mænjʊəlɪ]	*adv.*用手，手动地
traffic	['træfɪk]	*n.*流量，通信量
hop	[hɒp]	*v.*（鸟，蛙等）跳跃
brouter	[b'ru:tə]	*n.*网桥路由器
exact	[ɪg'zækt]	*adj.*精确的，准确的，原样的
gateway	['geɪtweɪ]	*n.*网关
configuration	[kənˌfɪgə'reɪʃn]	*n.*构造，结构，配置
default	[dɪ'fɔ:lt]	*n.*默认（值），缺省（值）
destine	['destɪn]	*vt.*注定，预定

Phrases

twisted pair	双绞线
fiber optic	光缆，光纤
active hub	有源集线器
thin Ethernet network	细线以太网，细缆以太网
token-ring network	令牌环网
logical bus topology	逻辑总线拓扑
logical ring topology	逻辑环形拓扑
physical star	物理星形
divide … into …	把……分为……
network layer	网络层
reference model	参考模型
routing table	路由表
mesh topology	网状拓扑
routable protocol	可路由协议
static router	静态路由器
dynamic router	动态路由器
packet switching network	分组交换网

Abbreviations

RJ (registered jack)	注册的插座
BNC (Bayonet Nut Connector)	刺刀螺母连接器，同轴电缆接插件
COAX (Coaxial Cable)	同轴电缆
UTP (Unshielded Twisted Paired)	非屏蔽双绞线
OSI (Open System Interconnect)	开放式系统互联
IPX (Internetwork Packet Exchange protocol)	互联网分组交换协议
SPX (Sequences Packet Exchange)	顺序分组交换
SMTP (Simple Mail Transfer Protocol)	简单邮件传输协议
GSNW (Gateway Service for NetWare)	NetWare 网关服务
PAD (Packet Assembler/Disassembler)	分组拆装器

Notes

[1] The hub is a hardware device that contains multiple, independent ports that match the cable type of the network.

本句中，that contains multiple, independent ports 是一个定语从句，修饰和限定 a hardware device。that match the cable type of the network 也是一个定语从句，修饰和限定 ports。

[2] If, for example, you have a thin Ethernet network with a cable segment longer than the prescribed maximum of 185 meters, you can install a repeater at some point in the segment to

strengthen the signals and increase the maximum segment length.

本句中，If 引导了一个条件状语从句。with a cable segment longer than the prescribed maximum of 185 meters 作定语，修饰和限定 a thin Ethernet network。to strengthen the signals and increase the maximum segment length 是动词不定式短语，作目的状语。

[3] The MAU does actively regenerate signals as it transmits data around the ring.

本句中，does 起强调作用，意思是“的确，确实”。

[4] Routers contain internal tables of information called routing tables that keep track of all known network addresses and possible paths throughout the internetwork, along with cost of reaching each network.

本句中，called routing tables 是过去分词短语，作定语，修饰和限定 internal tables of information。that keep track of all known network addresses and possible paths throughout the internetwork, along with cost of reaching each network 是一个定语从句，修饰和限定 routing tables。keep track of 的意思是“跟踪”。

[5] Routing tables are the means by which a router selects the fastest or nearest path to the next “hop” on the way to a data packet’s final destination.

本句中，by which a router selects the fastest or nearest path to the next “hop” on the way to a data packet’s final destination 是一个介词前置的定语从句，修饰和限定 the means。on the way to 的意思是“在……的途中”。

Text B

Networking Devices (2)

1. NIC (network interface card)

Network interface card, or NIC is a hardware card installed in a computer so it can communicate on a network. The network adapter provides one or more ports for the network cable to connect to, and it transmits and receives data onto the network cable.

1.1 Wireless LAN card

Every networked computer must also have a network adapter driver, which controls the network adapter. Each network adapter driver is configured to run with a certain type of network adapter.

1.2 Network card

Network adapters perform a variety of functions that are crucial to getting data to and from the computer over the network.

These functions are as follows:

1.2.1 Data encapsulation

The network adapter and its driver are responsible for building the frame around the data generated by the network layer protocol, in preparation for transmission. The network adapter also reads the contents of incoming frames and passes the data to the appropriate network layer protocol.

1.2.2 Signal encoding and decoding

The network adapter implements the physical layer encoding scheme that converts the binary data generated by the network layer-now encapsulated in the frame-into electrical voltages, light pulses, or whatever other signal type the network medium uses, and converts received signals to binary data for use by the network layer.

1.2.3 Transmission and reception

The primary function of the network adapter is to generate and transmit signals of the appropriate type over the network and to receive incoming signals. The nature of the signals depends on the network medium and the data link layer protocol. On a typical LAN, every computer receives all the packets transmitted over the network, and the network adapter examines the destination address in each packet to see if it is intended for that computer. If so, the network adapter passes the packet to the computer for processing by the next layer in the protocol stack; if not, the network adapter discards the packet.

1.2.4 Data buffering

Network adapters transmit and receive data one frame at a time, so they have built-in buffers that enable them to store data arriving either from the computer or from the network until a frame is complete and ready for processing.

1.2.5 Serial/parallel conversion

The communication between the computer and the network adapter runs in parallel, that is, either 16 or 32 bits at a time, depending on the bus the adapter uses. Network communications, however, are serial (running one bit at a time), so the network adapter is responsible for performing the conversion between the two types of transmission.

1.2.6 Media access control

The network adapter also implements the MAC mechanism that the data link layer protocol uses to regulate access to the network medium. The nature of the MAC mechanism depends on the protocol used.

2. ISDN (integrated service digital network) adapters

Integrated service digital network adapters can be used to send voice, data, audio, or video over standard telephone cabling. ISDN adapters must be connected directly to a digital telephone network. ISDN adapters are not actually modems since they neither modulate nor demodulate the digital ISDN signal.

Like standard modems, ISDN adapters are available both as internal devices that connect directly to a computer's expansion bus and as external devices that connect to one of a computer's serial or parallel ports. ISDN can provide data throughput rates from 56 Kbps to 1.544 Mbps (using a T1 carrier service).

ISDN hardware requires a NT (network termination) device, which converts network data signals into the signaling protocols used by ISDN. Sometimes, the NT interface is included, or integrated, with ISDN adapters and ISDN-compatible routers. In other cases, an NT device separate from the adapter or router must be implemented. ISDN works at the physical layer, data link layer, network layer, and transport layer of the OSI model.

3. WAP (wireless access point)

A wireless network adapter card with a transceiver, sometimes called an access point, broadcasts and receives signals to and from the surrounding computers and passes back and forth between the wireless computers and the cabled network.

Access points act as wireless hubs to link multiple wireless NICs into a single subnet. They also have at least one fixed Ethernet port to allow the wireless network to be bridged to a traditional wired Ethernet network.

4. Transceiver (media converter)

Transceiver, short for transmitter-receiver, is a device that both transmits and receives analog or digital signals. The term is used most frequently to describe the component in local area networks (LANs) that actually applies signals onto the network wire and detects signals passing through the wire. For many LANs, the transceiver is built into the network interface card (NIC). Some types of networks, however, require an external transceiver.

In Ethernet networks, a transceiver is also called a Medium Access Unit (MAU). Media converters interconnect different cable types (twisted pair, fiber, and thin or thick coax) within an existing network. They are often used to connect newer 100 Mbps, gigabit Ethernet, or ATM equipment to existing networks, which are generally 10BASE-T, 100BASE-T, or a mixture of both. They can also be used in pairs to insert a fiber segment into copper networks to increase cabling distances and enhance immunity to electromagnetic interference (EMI).

5. Firewalls

A firewall is a piece of hardware and/or software which functions in a networked environment to prevent some communications forbidden by the security policy, analogous to the function of firewalls in building construction(Figure 6-2).

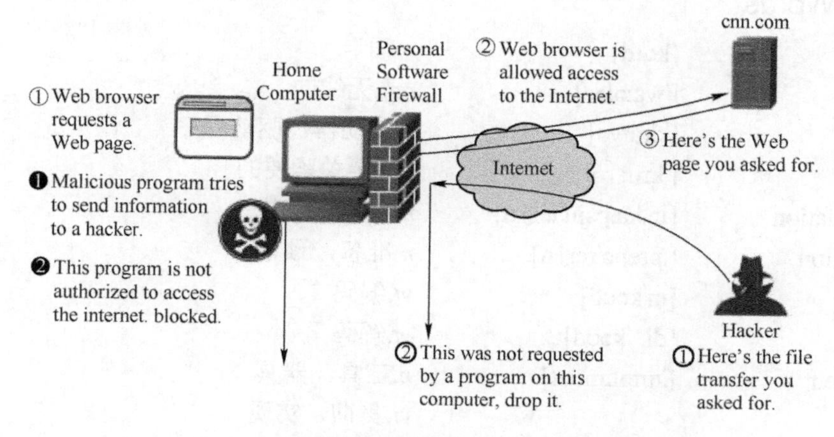

Figure 6-2 Firewall

A firewall has the basic task of controlling traffic between different zones of trust. Typical zones of trust include the Internet (a zone with no trust) and an internal network (a zone with high trust). The ultimate goal is to provide controlled connectivity between zones of differing trust levels through the enforcement of a security policy and connectivity model based on the least privilege principle.

There are three basic types of firewalls depending on:

- Whether the communication is being done between a single node and the network, or between two or more networks.
- Whether the communication is intercepted at the network layer, or at the application layer.
- Whether the communication state is being tracked at the firewall or not.

With regard to the scope of filtered communication exist these firewalls:

- Personal firewalls, a software application which normally filters traffic entering or leaving a single computer through the Internet.
- Network firewalls, normally running on a dedicated network device or computer positioned on the boundary of two or more networks or DMZs (demilitarized zones). Such a firewall filters all traffic entering or leaving the connected networks.

In reference to the layers where the traffic can be intercepted, there are two main categories of firewalls:

- Network layer firewalls, eg. ip tables.
- Application layer firewalls, eg. TCP Wrapper.

These network layer and application layer types of firewall may overlap, even though the personal firewall does not serve a network; indeed, single systems have implemented both together.

Lastly, depending on whether the firewalls track packet states, two additional categories of firewalls exist:

- Stateful firewalls.
- Stateless firewalls.

New Words

card	[kɑ:d]	*n.*插卡
wireless	['waɪələs]	*adj.*无线的
driver	['draɪvə]	*n.*驱动器，驱动程序
crucial	['kru:ʃəl]	*adj.*至关紧要的
encapsulation	[inˌkæpsju'leiʃn]	*n.*包装，封装
preparation	[ˌprepə'reɪʃn]	*n.*准备，预备
encode	[ɪn'kəʊd]	*vt.*编码
decode	[di: 'kəʊd]	*vt.*解码
implement	['ɪmplɪment]	*n.*工具，器具 *vt.*贯彻，实现
encapsulate	[ɪn'kæpsjuleɪt]	*v.*包装，封装
voltage	['vəʊltɪdʒ]	*n.*电压，伏特数
pulse	[pʌls]	*n.*脉冲

buffer	['bʌfə]	n.缓冲器
serial	['sɪərɪəl]	adj.串行的
conversion	[kən'vɜ:ʃn]	n.变换，转化
voice	[vɔɪs]	n.语音，声音
modem	['məʊdem]	n.调制解调器
modulate	['mɒdjuleɪt]	vt.（信号）调制
demodulate	[di:'mɒdjuleɪt]	vt.解调
expansion	[ɪk'spænʃn]	n.扩充，扩展
signaling	['sɪgnəlɪŋ]	n.发信号
surrounding	[sə'raʊndɪŋ]	n.环境 adj.周围的
gigabit	['gɪgəbɪt]	n.吉字节，吉比特
mixture	['mɪkstʃə]	n.混合，混合物
immunity	[ɪ'mju:nətɪ]	n.免疫性，免疫力
firewall	['faɪəwɔ:l]	n.防火墙
prevent	[prɪ'vent]	v.防止，预防
forbidden	[fə'bɪdn]	adj.禁止的，严禁的
enforcement	[ɪn'fɔ:smənt]	n.执行，强制
privilege	['prɪvəlɪdʒ]	n.特权，特别待遇 vt.给予……特权，特免
position	[pə'zɪʃn]	n.位置 vt.安置，决定……的位置
overlap	[ˌəʊvə'læp]	v.（与……）交叠，（与……）重叠

Phrases

telephone cabling	电话线
parallel port	并行端口
transport layer	传输层
access point	访问点，接入点
wired Ethernet network	有线以太网
analog signal	模拟信号
thick coax	粗同轴电缆
stateful firewall	状态防火墙，基于状态检测的防火墙

Abbreviations

NIC (Network Interface Card)	网络接口卡，网卡
ISDN (Integrated Service Digital Network)	综合业务数字网
Kbps (Kilobits per second)	千位/秒
Mbps (Megabits per second)	兆位/秒

NT (Network Termination)	网络终端
WAP (Wireless Access Point)	无线接入点，无线访问点
coax (coaxial cable)	同轴电缆
ATM (Asynchronous Transfer Mode)	异步传输模式
EMI (ElectroMagnetic Interference)	电磁干扰
DMZs (DeMilitarized Zones)	非军事区

Word Building

常用前缀列举如下：

1．counter-表示：反

action 作用 —— counteraction 反作用

clockwise 顺时针方向 —— counterclockwise 逆时针方向

2．dis-表示：否定、相反

order 秩序 —— disorder 混乱

charge 充电 —— discharge 放电

appear 出现 —— disappear 消失

3．en-表示：使……

large 大的 —— enlarge 扩大

close 紧的 —— enclose 封闭

danger 危险的 —— endanger 使危险

4．ex-表示：出自、向外

port 港口 —— export 出口

change 变化 —— exchange 交换

5．in-，表示：不、非、无

correct 正确的 —— incorrect 不正确的，错的

separable 可分开的 —— inseparable 不可分的

visible 看得见的 —— invisible 看不见的

注意，在 b、m、p 前加 im，在 l 前加 il。

possible 可能的 —— impossible 不可能的

logical 逻辑的 —— illogical 不合逻辑的

6．inter-表示：在……之间、在……之中

national 民族的 —— international 国际的

act 作用 —— interact 互相作用

7．kilo-表示：千

gram 克 —— kilogram 千克

metre 米—— kilometre 千米，公里

8．micro-表示：微、小

meter 米 —— micrometer 千分尺

switch 开关 —— microswitch 微型开关

Career Training

打 电 话

中国人打电话的习惯是直接找受话人，而不先报自己姓名。接电话时也是这样，第一句话往往是“您找哪位？”而不是先报自己姓名。而英语国家的人打电话的习惯却不是这样。他们往往是先向对方问好，然后报上自家姓名、电话号码等。如果主动打电话的一方想找另一方讲话，就直接说出他的名字。例如可以说：Is Mary there? Can I speak to Mary? 当接电话的人正是对方所要找的人时，可以说 Speaking，一般不说 Yes, I am。如果打电话的一方没有报出自己的姓名，接电话的人就可以询问对方是何人，如：May I know who is calling? 或 Who is speaking。但是不可以用 who are you 来询问。回答方也不能回答 I am Mary，而要用 This is Mary 来回答。

如果接电话的人不是对方要找的人，他可以让打电话的人稍等片刻，把需要的人找来。这时，接电话的人就可以说：Just a moment; Hold the line, please。如果要找的人不在，可以表示道歉，说那个人不在，也可以问一下对方是否需要留言。

和想要找的人通上话后，就和一般的对话没什么两样了。而结束通话的告别，也和普通告别无异。

打电话时的常用语如下：

Is Peter there?

Is Mr. White in?

Can I speak to Mary?

Is that you, Mike?

I would like to talk to Mary, please.

This is Mary speaking.

You are wanted on the phone.

It is for you.

Hold the line, please.

I will put you through.

Sorry, she is out.

Sorry, she is not available.

Would you like to leave a message?

Can I take a message for you?

You’ve got the wrong number.

There is nobody here by that name.

Would you please tell her to call back after 9:00?

Thank you for calling.

It is very kind of you.

Thanks, I will call her later.

软件水平考试试题解析

【真题再现】

Structured programming practices __A__ rise to Pascal, in which constructs were introduced to make programs more readable and better __B__. C provided a combination of assembly language and high-level structure to create a general-purpose language that could be used from system to __C__ programming. Next came object orientation, which is __D__ of a methodology and design philosophy than a language issue. This is __E__ by the addition of so called OO extensions to current languages, such as C.

供选择的答案

A：	①giving	②given	③gave	④gives
B：	①structure	②structured	③constructs	④structures
C：	①logic	②function	③flexible	④application
D：	①more	②little	③a matter	④important
E：	①evidence	②evidenced	③evidences	④evidencing

【答案】A：③　B：②　C：④　D：①　E：②

【试题解析】

A：这是一道时态考查题，本句应为一般过去时，故选③ 。give rise to 的意思是“引起，使发生”。

B：make programs better structured 的意思是“使程序的结构更合理”。故选②。

C：logic 的意思是“逻辑的”，function 的意思是“功能”，flexible 的意思是“易弯的，挠性的，有弹性的”，application 的意思是“应用”。与 system 相对的是 application，其他不合题意，故选④。

D：a matter of 的意思是“……的问题”，不合题意，more of A than B 的意思是“与其说是 B，不如说是 A”，故选①。

E：evidence 的意思是“证实，证明”。这里是一个被动句，因此要用过去分词 evidenced。故选②。

【参考译文】

结构化程序设计实践导致了 Pascal 语言的产生，在 Pascal 中引入的构造，提高了程序的可读性，提供了更好的程序结构。C 则是将汇编语言和高级编程语言结合起来而产生的一种通用语言，可用于从系统程序到应用程序的设计。接着出现了面向对象的技术，与其说它是一种语言，不如说是一种方法学和设计哲学。将所谓的 OO（面向对象）扩展附加到 C 这样的主流语言，就证明了这一点。

Exercises

[Ex. 1] 根据 **Text A** 回答以下问题。

1) What is a hub?

2) What does an active hub do?

3) What is a repeater?

4) What is the maximum segment length for a UTP cable on an Ethernet network?

5) What must a switch be able to do?

6) What is a bridge used to do? Where does it operate?

7) Where are routers most often used?

8) Why cannot routers compile routing tables from the information in the data packets they process?

9) What is a brouter? What does it do?

10) What is a gateway? Where does it operate?

[Ex. 2] 根据 Text B 回答以下问题。

1) What is Network Interface Card?

2) What is the primary function of the network adapter?

3) What can ISDN adapters be used to do?

4) What do access points act as?

5) What is transceiver?

6) What is a transceiver also called in Ethernet networks?

7) What is a firewall?

8) What is the ultimate goal of firewalls?

9) What is a personal firewall?

10) How many main categories of firewalls are there in reference to the layers where the traffic

can be intercepted?

[Ex. 3] 把下列句子翻译为中文。

1) What is the difference between an active hub and a passive hub?

2) The topology of the network may be ring, star or bus.

3) In order to build a LAN with more than two computers, it's necessary to use a multiport repeater.

4) With the bus topology, all workstations are directly connected to the main backbone that carries the data.

5) Connecting several thousands of computers to a LAN can in theory be done using a star, a ring, or a bus topology.

6) The computer receives the packet from the router and the process repeats as long as the computer is communicating with the external system.

7) A brouter is a network bridge combined with a router.

8) These routes tell the server how to get to specific hosts or networks outside by going through the default gateway.

9) Despite their limitations, organizations still need to run gateway-based E-mail filtering software.

10) As with the Ethernet switches, you can view the running configuration in order to do some sanity checking of the configuration on the terminal servers, if required.

[Ex. 4] 软件水平考试真题自测。

从供选择的答案中选出应填入下列英语文句中____内的正确答案，把编号写在答卷的对应栏内。

With circuit switching, a ____1____path is established between two stations for communication. Switching and transmission resources within the network are ____2____for the exclusive use of the circuit for the duration of the connection. The connection is ____3____: Once it is established, it appears to attached devices as if there were a direct connection. Packet switching was designed to provide a more efficient facility than circuit switching for ____4____data traffic. Each packet contains some portion of the user data plus control information needed for proper functioning of the network. A key distinguishing element of packet-switching networks is whether the internal operation is datagram or virtual circuit . With internal virtual circuits, a route is defined between two endpoints and all packets for that virtual circuit follow the ____5____route, With internal datagrams, each packet is treated independently, and packets intended for the same destination may follow different routes.

(1) A. unique	B. dedicated	C. nondedicated	D. independent
(2) A. discarded	B. abandoned	C. reserved	D. broken
(3) A. indistinct	B. direct	C. indirect	D. transparent
(4) A. casual	B. bursty	C. limited	D. abundant
(5) A. same	B. different	C. single	D. multiple

[Ex. 5] 听短文填空。

Exercises 6

Bus network is a topology or configuration for a local area ____1____in which all nodes are connected to a main communications line or bus. On a ____2____network, each ____3____monitors activity on the line. Messages are detected ____4____all nodes but are accepted only by the node(s) to which they are addressed. A malfunctioning node ceases to ____5____but does not disrupt operation, as it might on a ____6____network, in which ____7____are passed from one node to the next. To avoid ____8____that occur when two or more nodes try to use the line at the ____9____time, bus networks commonly rely on collision detection or ____10____passing to regulate traffic.

Reference Translation

组网设备

1. 集线器

使用星形拓扑结构的网络需要一个中央点来连接设备。这个装置被称为集中器，因为它合并了网络设备所有的电缆。集中器的基本形式是集线器。

集线器是一个硬件设备，它包含了与网络电缆类型相匹配的多个独立端口。虽然也有同轴电缆 BNC 和光纤 BNC 集线器，但连接带有 RJ45 端的 3 类或 5 类双绞线的集线器最常见。集线器为在设备之间传输数据提供了一种廉价的选择方案，但不具备任何形式的智能。

集线器可以分为主动式和被动式。主动式集线器在把信号发往目的地之前，会增强并重建输入的信号。而被动式集线器对所使用的信号不做任何加工。

1.1 以太网集线器

以太网集线器也被称为多端口中继器。中继器是放大通过它的信号以抵消衰减影响的设备。例如，如果你有一个用细线以太网，线缆长度超过规定的最高值（185 米），可以在该段的某些节点安装一个中继器以便增强信号，并增加最大长度。这种类型的中继器只有两个 BNC 连接器，并且如今已经很少见了。

在 UTP 以太网网络中使用的集线器也是中继器，但它们可以有许多 RJ45 端口，而不是只有两个 BNC 连接器。当数据通过其任何端口进入集线器时，集线器放大该信号并通过所有其他端口将信号传出。这使星形网络具有共享介质，即让每台计算机都有自己单独的电缆。集线器把网络上任何计算机发送给其他计算机的每一个数据包进行中继并放大信号。

以太网网络中，UTP 电缆最大网段长度为 100 米。一个段被定义为两台通信计算机之间的距离。但是，由于集线器也具有中继器的功能，每台计算机连接到集线器端口的电缆可长达 100 米，当一个集线器被插入到网络中时，允许的段长度可高达 200 米。

1.2　多站访问单元

多站访问单元（MAU）是一种用于令牌环网络的特殊类型的集线器。“集线器”一词通常与以太网相关，而 MAU 仅指令牌环网络。从外表看，MAU 像一个集线器。它连接到多个网络设备，每一个网络设备都使用单独的线缆。

与在物理星形网络中使用逻辑总线拓扑结构的集线器不同，MAU 在物理星形网络中采用了逻辑环形拓扑结构。

当 MAU 检测到连接问题时，环将发送信标。由于采用物理星形拓扑结构，MAU 可以很容易地检测到是哪个端口上存在问题，并关闭该端口，或“封闭”它。MAU 在环路中传输数据时主动再次生成信号。

2. 交换机

交换机是一种特殊类型的集线器，它比基本的、物理层的中继器和集线器更智能。交换机必须能够读取它收到的每个帧的 MAC 地址。这个信息让交换机把输入的数据帧只转发给计算机或被寻址的计算机，提升了网络速度并减少了拥堵。

交换机工作在 OSI 模型的物理层和数据链路层。

3. 桥接器

桥接器用以连接两个网段，允许每个网段上的计算机访问另一网段上的资源。它们也可用于将大型网络分成更小的片段。桥接器具有中继器的所有功能，但可以有更多的节点，而且由于对网络进行了分段，因而每段上竞争资源的计算机更少，从而提高了网络的性能。

桥接器也可以连接不同速度、不同拓扑结构或不同协议的网络。但不能把以太网段和令牌环段连接起来，因为它们使用了不同的网络标准。桥接器同时在物理层和数据链路层的 MAC 子层进行操作。桥接器读取各帧的 MAC 报头，以确定目的设备在桥的哪一侧，然后把信息传输到该设备所在的那个网段。

4. 路由器

路由器是从一个逻辑网络到另一个逻辑网络转发数据包的网络设备，以便延伸网络或对网络分段。路由器是大型网络互联时最常用的设备，这些网络使用 TCP/IP 来连接 TCP/IP 主机和局域网（LAN），以便使用租用专线访问因特网。

路由器在开放系统互连（OSI）参考模型的网络层（第 3 层）实现联网，以便在使用其逻辑地址（在使用 TCP/IP 的情况下，是目标主机上的 IP 地址）的网络间传输数据。因为路由器在 OSI 参考模型中工作的层次比桥接器更高，具有数据包路由能力更强、过滤功能更好及处理能力更强等特点，因而比桥接器花费更多。

路由器包含的内部信息表被称为路由表，路由表跟踪整个互联网络所有已知的网络地址和可能的路径，以及到达每个网络的成本。路由器根据可用的路径及其成本传送数据包，从而利用一个网状拓扑网络中存在的冗余路径。

因为路由器使用数据包的目的网络地址，只有配置的网络协议是 TCP/IP 或 IPX/SPX 这样的可路由的协议，它们才能工作。这与桥接器不同，桥接器与协议无关。路由表是路由器的心脏，没有它们，路由器就没有办法知道它接收到的数据包要发送到哪里。

与桥接器和交换机不同，路由器不能根据所处理的数据包中的信息来编译路由表。这是因为路由表所包含的信息比数据包中的更详细，并且路由器需要路由表中的信息来处理被激活后接收到的第一批数据包。路由器不能像桥接器那样把数据包转发到所有可能的目的地址。

静态路由器：必须手动配置其路由表与互联网上用到的所有网络地址和路径。

动态路由器：自动侦听网络通信流量以建立自己的路由表。

路由表是路由器选择通往数据包的最终目标的过程中，到达下一“跳”的最快或最近的路径的手段。这一选择过程通过使用路由度量来完成。

路由度量是确定数据包到达最终目的地的传输距离或时间的方法。路由度量有不同的形式。

路由器可以用于对大型网络分段并把局域网段连接到使用不同物理层和数据链路层标准的单一网干上。它们也可以用于把局域网连接到广域网。

5. 网桥路由器

网桥路由器是路由器和桥接器的组合。它是用于网络的一个特殊类型的设备，根据所转发的协议，网络可以由桥接器或路由器连接。网桥路由器是复杂的、相当昂贵的设备，因此很少采用。

一个网桥路由器几乎同时传送两种类型的流量：桥接流量和路由流量。对于桥接流量，该网桥路由器处理流量的方式与桥接器或交换机的方法一样，即根据数据包的物理地址转发数据。这使得桥接流量的传输速度相当快，但比通过桥接器直接传送要慢，因为网桥路由器必须确定该数据包是用桥接器还是用路由器来传输。

6. 网关

网关是用于连接使用不同协议的网络的设备。网关工作在 OSI 参考模型的网络层。为了与另一网络上的主机进行通信，IP 主机必须具有到目的地网络的路由配置。如果没有找到路由配置，主机将使用网关（默认 IP 路由器）把流量传输到目的主机。默认网关是 IP 向远程网络发送数据包的地方。如果没有指定默认网关，通信就被限制在本地网络。网关接收来自使用一种协议栈的网络数据，删除该协议栈并用另外网络使用的协议栈将其重新包装。

示例：

- 电子邮件网关，一个接收简单邮件传输协议（SMTP）电子邮件，把电子邮件转换成一个标准的 X.400 格式并转发到目的地的网关。
- Netware 网关服务（GSNW），使运行 Microsoft Windows NT server 或 Windows server 的机器为 Windows 客户端提供网关，使它们可以访问 Netware 服务器上的文件和打印资源。
- 在系统网络体系结构（SNA）主机和 TCP/IP 网络上的计算机之间的网关，如 Microsoft SNA 服务器所提供的一个网关。
- 一个分组拆装器（PAD），它提供局域网（LAN）与 X.25 分组交换网之间的连接。

Unit 7

Text A

Topologies

1. What is a topology?

A topology refers to the manner in which the cable is run to individual workstations on the network.[1] The dictionary defines topology as: the configurations formed by the connections between devices on a local area network (LAN) or between two or more LANs.

There are three basic network topologies (not counting variations thereon): the bus, the star, and the ring.

It is important to make a distinction between a topology and an architecture. A topology is concerned with the physical arrangement of the network components. In contrast, an architecture addresses the components themselves and how a system is structured (cable access methods, lower level protocols, topology, etc.). An example of an architecture is 10BASE-T Ethernet which typically uses the star topology.

2. What is a bus topology?

A bus topology connects each computer (node) to a single segment trunk. A trunk is a communication line, typically coaxial cable that is referred to as the bus. The signal travels from one end of the bus to the other. A terminator is required at each end to absorb the signal so it does not reflect back across the bus (Figure 7-1).

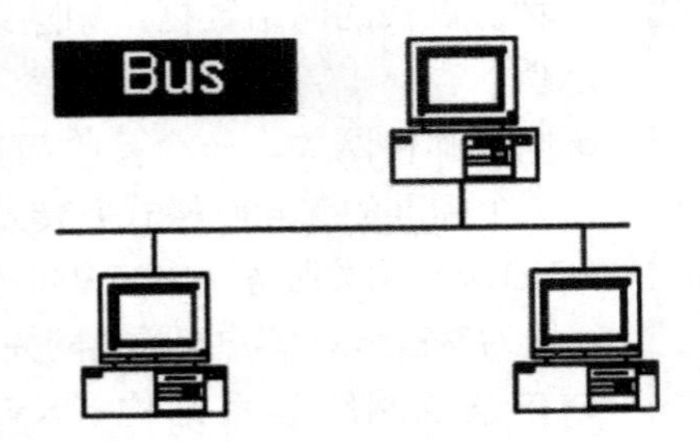

Figure 7-1 Bus topology

In a bus topology, signals are broadcast to all stations. Each computer checks the address on the signal (data frame) as it passes along the bus. If the signal's address matches that of the computer, the computer processes the signal. If the address doesn't match, the computer takes no action and the signal travels on down the bus.

Only one computer can talk on a network at a time. A media access method is used to handle the collisions that occur when two signals are placed on the wire at the same time.[2]

The bus topology is passive. In other words, the computers on the bus simply listen to a signal. They are not responsible for moving the signal along.

A bus topology is normally implemented with coaxial cable.

What are the advantages and disadvantages of the bus topology?

Advantages of bus topology:

- Easy to implement and extend.
- Well suited for temporary networks that must be set up in a hurry.
- Typically the least cheapest topology to implement.
- Failure of one station does not affect others.

Disadvantages of bus topology:

- Difficult to administer/troubleshoot.
- Limited cable length and number of stations.
- A cable break can disable the entire network; no redundancy.
- Maintenance costs may be higher in the long run.
- Performance degrades as additional computers are added.

3. What are the key features of a star topology?

All of the stations in a star topology are connected to a central unit called a hub (Figure 7-2). The hub offers a common connection for all stations on the network. Each station has its own direct cable connection to the hub. In most cases, this means more cable is required than for a bus topology. However, this makes adding or moving computers a relatively easy task. Simply plug them into a cable outlet on the wall.

If a cable is cut, it only affects the computer that was attached to it. This eliminates the single point of failure problem associated with the bus topology. (Unless, of course, the hub itself goes down.)

Star topologies are normally implemented using twisted pair cable, specifically unshielded twisted pair (UTP). The star topology is probably the most common form of network topology currently in use.

What are the advantages and disadvantages of a star topology?

Advantages of star topology:

- Easy to add new stations.
- Easy to monitor and troubleshoot.
- Can accommodate different wiring.

Disadvantages of star topology:

- Failure of hub cripples attached stations.
- More cable required.

4. What are the key features of a ring topology?

A ring topology consists of a set of stations connected serially by cable. In other words, it is a circle or ring of computers. There are no terminated ends to the cable. The signal travels around the circle in a clockwise direction (Figure 7-3).

Note that while this topology functions logically as ring, it is physically wired as a star. The central connector is not called a hub but a multistation access unit or MAU. (Don’t confuse a token ring MAU with a media adapter unit, which is actually a transceiver.[3])

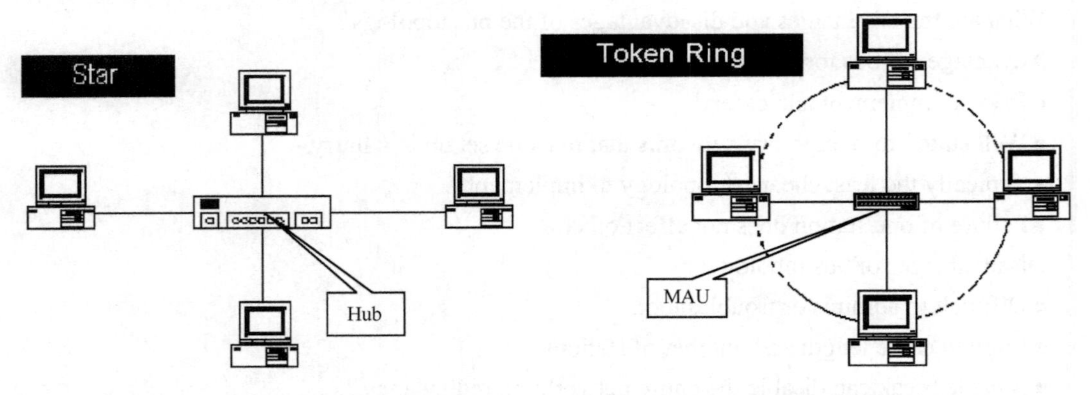

Figure 7-2 Star topology Figure 7-3 Ring topology

Under the ring concept, a signal is transferred sequentially via a "token" from one station to the next. When a station wants to transmit, it "grabs" the token, attaches data and an address to it, and then sends it around the ring. The token travels along the ring until it reaches the destination address. The receiving computer acknowledges receipt with a return message to the sender. The sender then releases the token for use by another computer.

Each station on the ring has equal access but only one station can talk at a time.

In contrast to the passive topology of the bus, the ring employs an active topology. Each station repeats or boosts the signal before passing it on to the next station.

Rings are normally implemented using twisted pair or fiber optic cable.

What are the advantages and disadvantages of a ring topology?

Advantages of ring topology:

- Growth of system has minimal impact on performance.
- All stations have equal access.

Disadvantages of ring topology:

- Most expensive topology.
- Failure of one computer may impact others.
- Complex.

5. Other FAQs

5.1 Why is a ring topology wired as a star?

A ring topology has the same outward appearance as a star. All the stations are individually connected to a central location. In the star topology, the device at the center is called a hub. In a ring topology, the center is called a MAU.

While they look the same, a closer examination reveals that the ring actually consists of a continuous circuit. Signals are passed along the circuit and accessed by stations in sequence. In a star topology the signal is split and sent out simultaneously to all stations.

5.2 What is a counterrotating ring?

A counterrotating ring is a ring topology that consists of two rings transmitting in opposite directions. The intent is to provide fault tolerance in the form of redundancy in the event of a cable failure. If one ring goes wrong, the data can flow across to the other path, thereby preserving the ring.[4]

5.3 Can you mix topologies?

Yes, you can mix various topologies on the same network.

One very common example is a large Ethernet network with multiple hubs. Usually the hubs are located on different floors in a building or perhaps outside in another building. Each hub is wired in the typical star configuration. However, the hubs are connected together along a bus, typically referred to as a backbone? The backbone between hubs might consist of fiber optic cable while the workstations are wired to each individual hub with UTP cable.

New Words

topology	[tə'pɒlədʒɪ]	*n.*拓扑
cable	['keɪbl]	*n.*电缆，缆，索；电报
individual	[ˌɪndɪ'vɪdʒuəl]	*n.*个人，个体 *adj.*个别的，单独的，个人的
workstation	['wɜ:ksteɪʃn]	*n.*工作站
variation	[ˌveəri'eɪʃn]	*n.*变种，变更，变异，变化
distinction	[dɪ'stɪŋkʃn]	*n.*区别，差别
arrangement	[ə'reɪndʒmənt]	*n.*排列，安排
Ethernet	['i:θənet]	*n.*以太网
node	[nəʊd]	*n.*节点
trunk	[trʌŋk]	*n.*干线，中继线
coaxial	[kəʊ'æksɪəl]	*adj.*同轴的，共轴的
terminator	['tɜ:mɪneɪtə]	*n.*终接器
absorb	[əb'sɔ:b]	*vt.*吸收
signal	['sɪgnəl]	*n.*信号 *adj.*作为信号的 *v.*发信号，用信号通知
reflect	[rɪ'flekt]	*v.*反射
broadcast	['brɔ:dkɑ:st]	*n.*广播，播音 *v.*广播，播送，播放
station	['steɪʃn]	*n.*站点
match	[mætʃ]	*n.& v.*匹配
media	['mi:dɪə]	*n.*媒体，介质
collision	[kə'lɪʒn]	*n.*碰撞，冲突
advantage	[əd'vɑ:ntɪdʒ]	*n.*优势，优点，有利条件
suit	[su:t]	*v.*合适，适合

failure	['feɪljə]	*n.*失败，失灵，故障
administer	[əd'mɪnɪstə]	*v.*管理，给予，执行
maintenance	['meɪntənəns]	*n.*维护，保持
degrade	[dɪ'greɪd]	*v.*退化，(使)降级，(使)退化
hub	[hʌb]	*n.*集线器
eliminate	[ɪ'lɪmɪneɪt]	*vt.*排除，消除，除去
cripple	['krɪpl]	*v.*使瘫痪
clockwise	['klɒkwaɪz]	*adj.*顺时针方向的 *adv.*顺时针方向地
adapter	[ə'dæptə]	*n.*适配器，转接器
transceiver	[træn'si:və]	*n.*收发器
token	['təʊkən]	*n.*令牌；标码
grab	[græb]	*v.*抓住
message	['mesɪdʒ]	*n.*消息，通讯，讯息 *vt.*通知
boost	[bu:st]	*v.*推进
complex	['kɒmpleks]	*adj.*复杂的，合成的，综合的 *n.*联合体
circuit	['sɜ:kɪt]	*n.*电路，一圈
counterrotating	['kaʊntərəʊ'teɪtɪŋ]	*adj.*反向旋转的
tolerance	['tɒlərəns]	*n.*容错
path	[pɑ:θ]	*n.*路经
backbone	['bækbəʊn]	*n.*中枢，骨干

Phrases

be concerned with	关心，关注，关于
star topology	星形拓扑
bus topology	总线拓扑
ring topology	环形拓扑
pass along	沿……走，路过
move along	向前移动
plug into	把插头插入，接通
twisted pair cable	双绞线
fiber optic cable	光纤电缆
together with	和，加之

Abbreviations

LAN (Local Area Network)	局域网，本地网
MAU (Multistation Access Unit)	多站访问（接入）单元

✍ Notes

[1] A topology refers to the manner in which the cable is run to individual workstations on the network.

本句中，in which the cable is run to individual workstations on the network 是一个介词前置的定语从句，修饰和限定 the manner。

英语中，由于意义或结构上的需要，通常使用介词+关系代词的结构。该结构可以引导限定性定语从句，也可以引导非限定性定语从句。该结构中的介词可以是 in、about、on、from、for、through、with、to、into、against 等，关系代词只能是 which 或 whom 不能是 that。请看下列：

Heat is a form of energy into which all other forms are convertible.

热是一种可以由其他所有能量转换来的能量。

There is only one problem about which they disagree.

现在他们只对一个问题持不同意见。

Water boils at 100 degrees centigrade, at which temperature it changes to gas.

水在摄氏一百度沸腾，达到该温度时水变成气体。

[2] A media access method is used to handle the collisions that occur when two signals are placed on the wire at the same time.

本句中，to handle the collisions that occur when two signals are placed on the wire at the same time 是一个动词不定式短语，作目的状语，修饰和限定 is used。在该短语中，when two signals are placed on the wire at the same time 是一个时间状语从句，修饰和限定 occur。that occur 是一个定语从句，修饰和限定 the collisions。

[3] Don't confuse a token ring MAU with a media adapter unit, which is actually a transceiver.

本句中，which is actually a transceiver 是一个非限定性定语从句，对 a media adapter unit 进行补充说明。confuse … with…的意思是"分不清……与……之间的不同，把……与……搞混淆了"。请看下列：

Don't confuse Austria with Australia.

不要把奥地利与澳大利亚弄混淆了。

[4] If one ring goes wrong, the data can flow across to the other path, thereby preserving the ring.

本句中，If one ring goes wrong 是一个条件状语；thereby preserving the ring 是一个现在分词短语，作结果状语。请看下列：

A number of new machines were introduced from abroad, thus resulting in an increase in production.

从国外引进了许多机器，因此产量增加了。

Text B

Text B
OSI

OSI

1. What is the OSI model?

The OSI model (open systems interconnection reference model) is a standard developed by the

ISO (International Standards Organization) to describe the flow of data on a computer network. The model takes into account the flow of data from the physical connections up to the end user application.

The OSI model defines a "layered" architecture in the form of a protocol stack. There are specific, discrete functions that take place at each layer of the protocol stack with lower level layers providing services to upper layers. When two systems communicate on the network, information is sent down through the protocol stack of one system, over the cable and then up through the protocol stack to the appropriate layer on the other system.

Data flow under the OSI model is organized into the following seven layers: physical, data link, network, transport, session, presentation, and application.

2. What is the purpose of the OSI model?

The OSI model (as well as IEEE 802 and NDIS) is established to standardize the design and construction of computer networks for developers and hardware manufacturers. Standards allow hardware and software components from a variety of different vendors to operate together. Without standards, everything would be vendor-specific and interoperability would suffer.

Almost all vendors implement standards in slightly different ways. This is normally done to take advantage of unique functionality in their product that just won't fit neatly into the model. However, if a vendor strays too far from the standards, they risk creating a "closed" or proprietary product that is not supported by the rest of the industry.

IBM mainframes use systems network architecture (SNA) that is a set of layered protocols like the OSI model. However, the SNA layers are not directly comparable to the OSI model layers. This makes interoperability between PC-based networks and IBM mainframes more difficult.

Exam Tip. Certification candidates need to understand the OSI and IEEE models because they provide a framework from which many of the technical concepts are taught. This understanding can be a big help when troubleshooting network problems in the real world.

3. What are the seven layers of the OSI model?

Application layer—user interface to network services such as file transfer/database access, terminal emulation.

Presentation layer—translates data format, data compression, redirector operations, network security.

Session layer—establish/maintain/terminate a connection; synchronization between user tasks with checkpoints, two way communication; name recognition.

Transport layer—flow control, ensures reliable delivery of packets in sequence with no losses or duplications (e.g., TCP & UDP).

Network layer—translation of logical and physical addresses, determines route from source to destination, traffic control (IP & IPX).

Data link layer—converts frames into bits (send side) and bits into frames (receive side), performs error checking, resends if no acknowledgement (e.g, the NIC driver).

Physical layer — transportation of raw (i.e., binary) data, defines cable & signaling specifications (e.g., the NIC & cable)

In real life things are never quite as precise as they are in models. For practical purposes you are generally only concerned with three "composite" layers: 1) the "lower layers" (physical and data link layers) that define the network architecture (e.g., Ethernet); 2) the "network" or "internetwork" layer (network and transport layers) where most of the addressing and logical transport functions are performed (e.g., TCP/IP); and the upper or "application" layer (session, presentation, and application) where high level networking functions take place (e.g., redirector and RPCs). The boundaries between these "composite" layers are separated by well-defined interfaces which on a Microsoft network are called NDIS and TDI.

Exam Tip. For purposes of certification testing, the mnemonic can be helpful in memorizing the names and order of the OSI layers.

All people seem to need data processing.

4. What is the function of the physical layer of the OSI model?

The physical layer of the OSI model establishes the physical characteristics of the network (e.g., the type of cable, connectors, length of cable, etc.). It also defines the electrical characteristics of the signals used to transmit the data (e.g. signal voltage swing, duration of voltages, etc.). The physical layer transmits the binary data (bits) as electrical or optical signals depending on the medium.

A repeater functions at the Physical Layer.

5. What is the function of the data link layer of the OSI model?

The data link layer defines how the signal will be placed on or taken off the NIC. It is here that the data frames are broken down into individual bits that can be translated into electronic signals and sent over the network. On the receiving side, the bits are reassembled into frames for processing by upper levels.

Error detection and correction is also performed at the data link layer. If an acknowledgement is expected and not received, the frame will be resent. Corrupt data is also identified at the data link layer.

Because the data link layer is very complex, it is sometimes divided into sublayers (as defined by the IEEE 802 model). The lower sublayer, called the MAC sublayer, provides network access. The upper sublayer, called the LLC sublayer, is concerned with sending and receiving packets and error checking.

A bridge functions at the data link layer.

6. What is the function of the network layer of the OSI model?

The network layer is primarily concerned addressing and routing. Logical addresses (e.g., an IP address) are translated into physical addresses (i.e., the MAC address) for transmission at the network layer. On the receiving side, the translation process is reversed.

It is at the network layer that the route from the source to destination computer is determined. Routes are determined based on packet addresses and network conditions. Traffic control measures are also implemented at the network layer.

Common protocols that operate at this level include IP, IPX, and X.25 (frame relay).

A router functions at the Network layer.

7.What is the function of the transport layer of the OSI model?

The transport layer is responsible for flow control and ensuring messages are delivered error free.

On the sending side, messages are packaged for efficient transmission at the transport layer and assigned a tracking number so they can be reassembled in proper order. On the receiving side, the packets are reassembled, checked for errors and acknowledged. The transport layer performs error handling in that it ensures all data is received in the proper sequence and without errors. If there are errors, the data is retransmitted.

Common protocols that operate at the transport layer include TCP, UDP, SPX, and NetBEUI.

8. What is the function of the session layer of the OSI model?

The session layer is responsible for establishing, maintaining, and terminating a connection called a "session". A session is an exchange of messages between computers (a dialog). Managing the session involves synchronization of user tasks and dialog control (e.g., who transmits and for how long). Synchronization involves the use of checkpoints in the data stream. In the event of a failure, only the data from the last checkpoint has to be resent.

Logon, name recognition and security functions take place at the session layer. Winsock and NetBIOS are usually shown as functioning at the session layer.

9. What is the function of the presentation layer of the OSI model?

The presentation layer is responsible for data translation (formatting), compression, and encryption.

The presentation layer is primarily concerned with translation, interpreting and converting the data from various formats. For example, EBCIDIC characters might be converted into ASCII. It is also where data is compressed for transmission and uncompressed on receipt. Encryption techniques are implemented at the presentation layer.

The redirector operates at the presentation layer by redirecting I/O operations across the network.

10. What is the function of the application layer of the OSI model?

The application layer provides the operating system with direct access to network services. It serves as the interface between the user and the network by providing services that directly support user applications. Note that a user-based application like Excel does not operate at the application layer. Instead, the application layer provides an interface so that processes such as Excel or Word that are running on the local machine can get access to network services (e.g., retrieving a file from a network server).

New Words

layered	['leɪəd]	*adj*.分层的
stack	[stæk]	*n*.堆，堆栈 *v*.堆叠

appropriate	[ə'prəʊprɪət]	*adj*.适当的
session	['seʃn]	*n*.会话，对话
presentation	[ˌprezn'teɪʃn]	*n*.表示，表达，介绍，陈述
vendor	['vendə]	*n*.卖主
interoperability	['ɪntərɒpərə'bɪlətɪ]	*n*.互通性，互用性，互操作性
risk	[rɪsk]	*vt*.冒……的危险 *n*.冒险，风险
proprietary	[prə'praɪətrɪ]	*adj*.私有的，私有财产的
mainframe	['meɪnfreɪm]	*n*.主机，大型计算机
certification	[ˌsɜ:tɪfɪ'keɪʃn]	*n*.证明，认证
framework	['freɪmwɜ:k]	*n*.构架，框架，结构
emulation	[ˌemjʊ'leɪʃn]	*n*.仿真，模拟
compression	[kəm'preʃn]	*n*.压缩
synchronization	[ˌsɪŋkrənaɪ'zeɪʃn]	*n*.同步
recognition	[ˌrekəg'nɪʃn]	*n*.识别
packet	['pækɪt]	*n*.包，信息包 *v*.包装
sequence	['si:kwəns]	*n*.次序，顺序，序列
acknowledgement	[ək'nɒlɪdʒmənt]	*n*.确认，承认
binary	['baɪnərɪ]	*adj*.二进制的，二进位的
composite	['kɒmpəzɪt]	*adj*.合成的，复合的 *n*.合成物
boundary	['baʊndrɪ]	*n*.边界，分界线，界限，范围
responsible	[rɪ'spɒnsəbl]	*adj*.负责的，可靠的，可依赖的
reassemble	[ˌri:ə'sembl]	*vt*.重新装配，重新集合
encryption	[ɪn'krɪpʃn]	*n*.加密，编密码
server	['sɜ:və]	*n*.服务器

Phrases

data link	数据链路
send side	发送方
receive side	接收方
error checking	差错校验
raw data	原始数据，未加工的数据
data stream	数据流

Abbreviations

IEEE (Institute for Electrical and Electronics Engineers)	电气和电子工程师学会
NDIS (Network Driver Interface Specification)	网络驱动器接口标准

SNA (System Network Architecture) 系统网络体系
UDP (User Datagram Protocol) 用户数据报协议
TCP (Transmission Control Protocol) 传输控制协议
TDI (Trandport Driver Interface) 传输驱动程序接口
ASCII (American Standard Code for Information Interchange) 美国信息交换标准码
IPC (InterProcess Communication) 内部进程通信
RPC (Remote Procedure Call) 远程过程调用

Word Building

常用前缀列举如下：

1．milli-表示：毫、千分之一

volt 伏 —— millivolt 毫伏

liter 升 —— milliliter 毫升

gram 克 —— milligram 毫克

2．mis-表示：误、错、坏

fortune 运气 —— misfortune 不幸

manage 管理 —— mismanage 对……管理不当

lead 引导 —— mislead 误导

understand 理解 —— misunderstand 误解

3．mono-表示：单、一

plane 飞机 —— monoplane 单翼飞机

rail 铁道 —— monorail 单轨铁道

tone 音调 —— monotone 单音，单调

4．multi-表示：多

colour 颜色 —— multicolour 多色的，五彩缤纷的

program 程序 —— multiprogram 多程序

form 形式 —— multiform 多形式的，多样的

5．non-表示：非、不、无

stop 停 —— nonstop 直达，中途不停

metal 金属 —— nonmetal 非金属

conductor 导体 —— nonconductor 绝缘体

6．over-表示：过分、在……上面；超过；压倒、额外

production 生产 —— overproduction 生产过剩

load 载 —— overload 过载

head 头 —— overhead 在头顶上的，在高处的

run 流 —— overrun 溢出，超越

7．poly-表示：多、复、聚

crystal 晶体 —— polycrystal 多晶体

technical 技术的，工艺的 —— polytechnical 多工艺的
atomic 原子的 —— polyatomic 多原子的
8．post-表示：后
war 战争 —— postwar 战后的
liberation 解放 —— post-liberation 解放后的
9．pre-表示：预先、在前
heat 加热 —— preheat 预热
condition 条件 —— precondition 前提，先决条件
pay 付款 —— prepay 预付，提前付
10．re-表示：再、重新
run 运行 —— rerun 重新运行
write 写 —— rewrite 改写
set 设置 —— reset 重新设置
print 打印 —— reprint 重新打印

Career Training

约　见

约见是人们日常交际中重要的活动之一。当我们需要和别人一起参加某项活动或拜访他人时，应当征求别人的意见，提前约定。

在英美国家，预约尤其重要，不管做什么事情，他们都需要提前与别人约好，不会唐突造访。如果不事先通知就贸然造访，是相当失礼的行为。一般来讲，人们都用电话来约定见面，方便又快捷。

约见时应该注意以下3点。

1）正确介绍自己的姓名，并正确获知对方姓名。

2）要询问对方希望与谁晤面，并且把信息正确地传达给对方。

3）要传达或询问约定的时间与场所。

如果掌握了上述3条原则，再牢记一些基本句型，你就能流利地用英语预约了。

对于商业预约，约好时间地点后，在见面之前我们可以用“Can I ask you what you want to discuss?”，事先了解待讨论的内容、关系、背景等，这将有助于做一些准备工作。

如果对方没有时间，可以用“How about...?”的句型来调整，以约定适当的见面时间。告诉对方你的空闲时间，可以让对方选择方便的时间。

当你确实没空，不能答应对方的约定时，最好说“I'm afraid I can't...”，这样会使人觉得你非常客气。若不小心唐突地说出“I'm too busy to meet you”，可能会使对方感到受轻视。

对于不认识的人或推销人员的求见电话，你认为无须见面时，可以用这句话，或明确地用“It's probably not necessary to meet.”来加以谢绝。

预约时的常用语如下：

I'd like to make an appointment with Mr. Jones.
I don't think we have to meet on this subject.
I was wondering if we could arrange a meeting to discuss the new product.
May I have your name address/ phone number please?
Will you be free tomorrow?
Do you have time this afternoon?
How about tomorrow morning?
When / Where shall we meet?
May I arrange the time and the place, please?
Could we get together at L.A. Restaurant?
What time is convenient for you?
Can you manage five (o'clock)?
Will five o'clock be all right?
What about five (o'clock)?
Could you spare me about half an hour?
Could we meet at 4: 30?
Let's make it 4: 30.
All right. See you then.
I'm afraid I have no time then.
Sorry, I won't be free then. But I'll be free tomorrow.

软件水平考试试题解析

【真题再现】

从供选择的答案中选出应填入下列英语文句中____内的正确答案，把编号写在答卷的对应栏内。

Although parallel server hardware has been available for some time, commercially available parallel versions of database __A__ are just now arriving from __B__ database makers to take advantage of the hardware's speed.

Parallel hardware and database __C__ are probably overkill for some applications, such as small departmental systems, analysts say. But users at large sites now installing parallel systems say they can __D__ improve database response __E__ for large decision-support, order-entry, and data warehouse applications.

供选择的答案

A：①hardware ②software ③firmware ④netware
B：①leading ②lead ③leader ④leadering
C：①combination ②collection ③combinations ④collections
D：①differently ②drastically ③important ④good
E：①action ②condition ③space ④time

【答案】A：② B：① C：③ D：② E：④

【试题解析】

A：parallel server hardware 的意思是“平行服务器硬件”，database software 的意思是“数据库软件”，firmware 的意思是“固件”，netware 的意思是“网件”。根据句意，应选②。

B：此处应填一个形容词，修饰 database makers。leading 是个形容词，意思是“最重要的；主要的”，故选①。lead 是动词，意思是“领导，带领”；leader 是名词，意思是“领导者，领导，领袖”；leadering 不存在，是出题人杜撰的。

C：collection 的意思是“收集，收取”，combination 的意思是“结合，组合”。根据语句结构，此处应该用复数形式，故选 ③。

D：此处应填一个副词，因此只能在 differently 和 drastically 中进行选择，differently 的意思是“不同地”，drastically 的意思是“激烈地，彻底地”，根据句子的意思，选②。

E：database response time 的意思是“数据库的响应时间”。其他搭配不合理，故选④。

【参考译文】

尽管并行服务器硬件已经出现一段时间了，但能发挥硬件速度的商用数据库软件的并行版本，目前才刚刚可以由领先的数据库厂商提供。

虽然据分析家说，并行硬件和数据库的结合可能会扼杀一些应用软件，像小型部分系统；但是，目前那些安装了并行系统的大型站点的用户说，它们彻底地优化了大型决策支持系统、订单管理系统和数据仓库应用软件的数据库响应时间。

Exercises

[Ex. 1] 根据 Text A 回答以下问题。

1) What does topology refer to?

2) What is the definition of topology given in the dictionary?

3) How many basic network topologies are there mentioned in the text? What are they?

4) What is a bus topology?

5) What are the advantages and disadvantages of the bus topology?

6) How are star topologies normally implemented?

7) What are the advantages and disadvantages of a star topology?

8) What does a ring topology consist of?

9) What are the advantages and disadvantages of a ring topology?

10) Can you mix topologies? What is the example given in the text?

[Ex. 2] 根据 Text B 填空。

1) The OSI model is ____________ developed by the ISO to__________________ on a computer network.

2) Data flow under the OSI model is organized into seven layers: ____________, ____________, ____________, ____________, ____________, ____________ and ____________.

3) The OSI model is established to ____________________________ of computer networks for developers and ______________.

4) IBM mainframes use ____________________________ that is a set of layered protocols like ______________.

5) The mnemonic which can be helpful in memorizing the names and order of the OSI layers are ______________________________________.

6) The Physical Layer of the OSI model establishes ____________________________ of the network (e.g., the type of cable, connectors, length of cable, etc.). It also defines ____________________________ of the signals used to transmit the data.

7) Because the data link layer is very complex, it is sometimes divided into ____________.

8) It is _________________ that the route from the source to ____________ is determined.

9) The transport layer is responsible for ____________ and ensuring messages are delivered ____________. The session layer is responsible for ______________, ______________, and terminating a connection called a "session".

10) The presentation layer is primarily concerned with ______________, ______________, and ______________ from various formats. The application layer serves as the interface between ______________and ______________, by providing services that directly support user applications.

[Ex. 3] 把下列句子翻译为中文。

1) Star network is a LAN in which each device (node) is connected to a central computer in a star-shaped configuration.

2) Cable modem sends and receives data through a coaxial cable television network instead of telephone lines.

3) There is no appreciable distinction between the two designs.

4) Adapter enables a personal computer to use a peripheral device.

5) Bus mouse attaches to the computer's bus through a special card or port rather than through a serial port.

6) A mouse, a keyboard, and a printer might all be connected to a computer with cables.

7) Network nodes are linked by coaxial cable, by fiber optic cable, or by twisted-pair wiring.

8) Token ring network is formed in a ring topology that uses token passing as a means of regulating traffic on the line.

9) On a token ring network, a token governing the right to transmit is passed from one station to the next in a physical circle.

10) Token bus network is formed in a bus topology that uses token passing as a means of regulating traffic on the line.

[Ex. 4] 软件水平考试真题自测。

从供选择的答案中选出应填入下列英语文句中____内的正确答案，把编号写在答卷的对应栏内。

Software quality assurance is now an __A__ sub-discipline of software engineering. As Buckly and Oston point out, __B__ software quality assurance is likely to lead to an ultimate __C__ of software costs. However, the major hurdle in the path of software management in this area is the lack of __D__ software standards. The development of accepted and generally applicable standards should be one of the principal goals of __E__ in software engineering.

供选择的答案：

A：①emerging ②emergent ③engaging ④evolve

B：①effective ②effortless ③light ④week

C：①balance ②growth ③production ④reduction

D：①usable ②usage ③useless ④useness

E：①management ②planning ③production ④research

[Ex. 5] 听短文填空。

Exercises 7

LAN stands for local area network. A LAN is a group of computers and devices that are in a ____1____ location. The devices connect to the LAN with an Ethernet ____2____ or through WiFi. Your home may have a LAN. If your PC, tablet, smart TV, and wireless ____3____ connect through your WiFi, these connected devices are part of your LAN. Only devices that you authorize have access to your LAN.

LANs come in many ____4____. A group of devices connected through a home internet connection is a LAN. Small businesses have LANs that connect a dozen or a hundred computers with printers and ____5____ storage. The largest LANs are controlled by a server that stores files, ____6____ data between devices, and directs files to printers and scanners.

A LAN ____7____ from other types of computer networks in that the devices connected to the

LAN are in the same ____8____ such as a home, school, or office. These computers, printers, scanners, and other devices connect to a ____9____ with an Ethernet cable or through a wireless router and a WiFi access point. Multiple LANs can be connected over a ____10____ line or radio wave.

Reference Translation

拓　　扑

1. 什么是拓扑？

拓扑是指网络中各个独立工作站的电缆布局方式。词典中拓扑的定义是局域网上设备之间或多个局域网之间的连接形成的配置。

有3种基本的网络拓扑结构（不考虑它们的变种）：总线、星形及环形。

区分拓扑与体系结构之间的差别是重要的。拓扑是网络设备相互连接的物理布局。相反，体系结构是部件自身的连接，以及系统的构成（电缆连通方法、低层协议、拓扑等）。专用于星形拓扑的10BASE-T以太网就是体系结构的一个例子。

2. 什么是总线拓扑？

总线拓扑把每台计算机（节点）连接到一个单段干线。干线就是一个通信线，通常使用同轴电缆，被称作总线。信号从总线的一端传输到另一端。每一端点需要一个终接器来吸收信号，以防止回波进入总线。

在总线拓扑中，信号被广播给全部站点。每台计算机检查通过总线传输的信号的地址（数据帧）。如果数据的地址与该计算机匹配，计算机就处理该信号。如果数据的地址与该计算机不匹配，计算机就不动作，信号沿总线继续向下传输。

一次只能有一台计算机发出信号。媒体访问模式用于处理当两个信号同时在总线上传输时产生的碰撞。

总线拓扑是被动的。换言之，总线上的计算机只能简单地听信号，而不负责向前传输信号。

总线拓扑通常由同轴电缆实现。

总线拓扑有何优点与缺点？

总线拓扑的优点：

- 容易实现和扩展。
- 非常适合必须马上建设的临时网络。
- 实现成本特别低。
- 一个站点的故障不会影响其他站点。

总线拓扑的缺点：

- 难于管理和排查故障。
- 电缆长度和站点数受到限制。
- 电缆断开可能会影响整个网络，没有冗余。
- 长期维护成本比较高。

- 随着计算机的增加，性能降低。

3. 星形拓扑的主要特点是什么？

星形拓扑中的全部站点都被连接到叫作“集线器”的中央单元。

集线器为网络上的每个站点提供公共连接。每个站点用自己的直联电缆连接到集线器上。在大多数情况下，这意味着需要比总线拓扑更多的电缆。但是，这使增加或减少计算机变得相当容易。只要把计算机插到墙上的电缆出口就行了。

如果电缆断开，只会影响到它所连接的计算机。这就消除了总线拓扑中单点故障影响整个网络的问题（当然除非集线器自己出问题了）。

星形拓扑通常由双绞线实现，特别是非屏蔽双绞线。星形拓扑在现在所用的网络拓扑中大概是最常见的。

星形拓扑有何优点与缺点？

星形拓扑的优点：

- 容易增加新站点。
- 容易监控和排查故障。
- 可以提供不同的接线。

星形拓扑的缺点：

- 集线器的故障会造成整个网络瘫痪。
- 需要更多的电缆。

4. 环形拓扑的主要特点是什么？

环形拓扑由电缆串联起来的一套站点构成。换言之，它是由计算机组成的一个环或一个圈。电缆没有终点。信号沿这个环以顺时针方向传输。

注意，尽管环形拓扑的运行在逻辑上是环形的，但其物理连线是星形的。中央控制单元不是集线器而是“多站访问单元”（不要把令牌环 MAU 与媒体适配器单元相混淆，后者实际上是收发器）。

在环形拓扑中，信号经“令牌”由一个站点传输到另一个站点。当一个站点要发送信号时，它“夺得”令牌，给其“绑缚”数据及目的地址，然后绕环发送。令牌沿环路传播直到到达目的地址。接收信息的计算机给发送信息的站点发送回复信息告知已收到信息。然后，发送信息的站点释放令牌给其他计算机使用。

环形拓扑中的每个站点都是平等的，但一次只能有一个站点可以发送信号。

与被动的总线拓扑相比，环形使用主动拓扑。每个站点重复或引导信号后，才把信号传递到下一站点。

环形拓扑通常使用双绞线或光纤电缆。

环形拓扑有何优点与缺点？

环形拓扑的优点：

- 系统的增长会最小限度地影响性能。
- 全部站点都可以平等访问。

环形拓扑的缺点：

- 最昂贵的拓扑。
- 一台计算机的故障会影响其他计算机。

- 复杂。

5. 其他常见问题

5.1　为什么环形拓扑的连线为星形？

环形拓扑外表看起来与星形相同。全部的站点都独立地连接到中央位置。在星形拓扑中，中央设备是集线器，而环形拓扑的中央设备是MAU。

虽然它们看上去相同，但仔细研究可以发现：环形实际上由连续的线路组成，信号通过这些线路并有序地被一系列站点访问；在星形拓扑中，信息被分开并同时发送到全部站点。

5.2　什么是反向环？

反向环是由两个相反方向的环构成的环形拓扑。其目的是在电缆发生故障时，以冗余形式提供容错。如一个环出错，数据可以流过另外的路径，从而保护环路。

5.3　可以混合拓扑吗？

可以的，可以在一个网络中混合多种拓扑结构。

一个常见的例子是带多个集线器的大型以太网。通常这些集线器位于大楼的不同楼层，或者也许在其他楼内。每个集线器使用典型的星形拓扑结构连接起来。但是，这些集线器都连接到一个总线中，该总线通常被称为骨干网（也称“主干网”）。集线器之间的骨干网可以由光纤组成，而工作站由非屏蔽双绞线连接到每个集线器上。

Unit 8

Text A

Network Security

Text A
Network Security

Network security consists of the policies and practices adopted to prevent and monitor unauthorized access, misuse, modification, or denial of a computer network and network-accessible resources. Network security involves the authorization of access to data in a network, which is controlled by the network administrator.[1] Users choose or are assigned an ID and a password or other authentication information that allows them access to information and programs within their authority. Network security covers a variety of computer networks, both public and private, that are used in common everyday jobs; conducting transactions and communications among businesses, government agencies and individuals. Networks can be private, such as within a company, and others which might be open to public access. Network security is involved in organizations, enterprises, and other types of institutions. It does as its title explains: it secures the network, as well as protecting and overseeing operations being done. The most common and simple way of protecting a network resource is by assigning it a unique name and a corresponding password.

Network security starts with authentication, commonly with a username and a password. Since this requires just one detailed authentication, this is sometimes termed one-factor authentication. With two-factor authentication, something the user "has" is also used (e.g., a security token , an ATM card, or a mobile phone); and with three-factor authentication, something the user "is" also used (e.g., a fingerprint or retinal scan).

Once authenticated, a firewall enforces access policies such as what services are allowed to be accessed by the network users. Though effective to prevent unauthorized access, this component may fail to check potentially harmful content, such as computer worms or Trojan horses being transmitted over the network.[2] Antivirus software or an intrusion prevention system (IPS) help detect and inhibit the action of such malware. An anomaly-based intrusion detection system may also monitor the network like wireshark traffic and may be logged for audit purposes and for later high-level analysis. Newer systems combining unsupervised machine learning with full network traffic analysis can detect active network attackers from malicious insiders or targeted external attackers that have compromised a user machine or account.[3]

Communication between two hosts using a network may be encrypted to maintain privacy.

Honeypots, essentially decoy network-accessible resources, may be deployed in a network as surveillance and early warning tools, as the honeypots are not normally accessed for legitimate purposes. Techniques used by the attackers that attempt to compromise these decoy resources are studied during and after an attack to keep an eye on new exploitation techniques.[4] Such analysis may be used to further tighten security of the actual network being protected by the honeypot. A honeypot can also direct an attacker's attention away from legitimate servers. A honeypot encourages attackers to spend their time and energy on the decoy server while distracting their attention from the data on the real server. Similar to a honeypot, a honeynet is a network set up with intentional vulnerabilities. Its purpose is also to invite attacks so that the attacker's methods can be studied and that information can be used to increase network security. A honeynet typically contains one or more honeypots.

With all of the vital personal and business data being shared on computer networks every day, security has become one of the most essential aspects of networking. No one recipe to fully safeguard networks against intruders exists. Network security technology improves and evolves over time as the methods for both attack and defense grow more sophisticated.

1. Physical network security

The most basic but often overlooked element of network security involves keeping hardware protected from theft or physical intrusion.

Corporations spend large sums of money to lock their network servers, network switches and other core network components in well-guarded facilities. While these measures aren't practical for homeowners, households should still keep their broadband routers in private locations, away from nosy neighbors and house guests.

The widespread use of mobile devices makes physical security much more important. Small gadgets are especially easy to leave behind at travel stops or to have fall out of pockets.

Finally, stay in visual contact with a phone when loaning it to someone else: a malicious person can steal personal data, install monitoring software, or otherwise "hack" phones in just a few minutes when left unattended.

2. Password protection

Passwords are an extremely effective system for improving network security if applied properly. Unfortunately, some don't take password management seriously and insist on using bad, weak (meaning, easy to guess) passwords like "123456" on their systems and networks.

Following just a few common-sense best practices in password management will greatly improve the security protection on a computer network:

- Set strong passwords, or passcodes, on all devices that join the network.
- Change the default administrator password of network routers.
- Do not share passwords with others more often than necessary; set up guest network access for friends and visitors if possible; change passwords when they may have become too widely known.

3. Spyware

Even without physical access to the devices or knowing any network passwords, illicit programs called spyware can infect computers and networks, typically by visiting Web sites. Much spyware exists on the Internet. Some spyware monitors a person's computer usage and Web browsing habits and reports this information back to corporations, who use it to create more targeted advertising. Other spyware attempts to steal personal data. One of the most dangerous forms of spyware, keylogger software, captures and sends the history of all keyboard keystrokes a person makes, ideal for capturing passwords and credit card numbers. All spyware on a computer attempts to function without the knowledge of people using it, thereby posing a substantial security risk.[5]

Because spyware is notoriously difficult to detect and remove, security experts recommend installing and running reputable anti-spyware software on computer networks.

4. Online privacy

Personal stalkers, identity thieves and perhaps even government agencies monitor people's online habits and movements well beyond the scope of basic spyware. WiFi hotspot usage from commuter trains and automobiles reveal a person's location, for example. Even in the virtual world, much about a person's identity can be tracked online through the IP addresses of their networks and their social network activities.

Techniques to protect a person's privacy online include anonymous Web proxy servers, although maintaining full privacy online is not fully achievable through today's technologies.

New Words

policy	['pɒləsɪ]	*n.*政策，方针
practice	['præktɪs]	*n.*实行，实践，实际，惯例，习惯
unauthorized	[ʌn'ɔ:θəraɪzd]	*adj.*未被授权的，未经认可的
access	['ækses]	*n.*访问，入门，接入 *vt.*存取
denial	[dɪ'naɪəl]	*n.*否认，否定，拒绝
resource	[rɪ'sɔ:s]	*n.*资源
authorization	[ˌɔ:θəraɪ'zeɪʃn]	*n.*授权，认可
data	['deɪtə]	*n.*数据
control	[kən'trəʊl]	*n.&vt.*控制，支配，管理，操纵
administrator	[əd'mɪnɪstreɪtə]	*n.*管理员
password	['pɑ:swɜ:d]	*n.*密码，口令
authenticate	[ɔ:'θentɪkeɪt]	*v.*鉴别
authority	[ɔ:'θɒrətɪ]	*n.*权威，威信
cover	['kʌvə]	*vt.*包括，包含，适用
communication	[kəˌmju:nɪ'keɪʃn]	*n.*通信
institution	[ˌɪnstɪ'tju:ʃn]	*n.*公共机构，协会
explain	[ɪk'spleɪn]	*v.*解释，说明
oversee	[ˌəʊvə'si:]	*v.*监视，检查
username	['ju:zəneɪm]	*n.*用户名
fingerprint	['fɪŋ gəprɪnt]	*n.*指纹，手印

vt.采指纹

component [kəm'pəʊnənt] n.成分；组件

adj.组成的，构成的

Trojan (horse) ['trəʊdʒən] n.特洛伊木马

transmit [træns'mɪt] vt.传输，传播

inhibit [ɪn'hɪbɪt] v.抑制，约束

malware ['mælweə] n.恶意软件

anomaly [ə'nɒməlɪ] n.不异常的

unsupervised [ˌʌn'sju:pəvaɪzd] adj.无人监督的，无人管理的

attacker [ə'tækə] n.攻击者

malicious [mə'lɪʃəs] adj.怀有恶意的，恶毒的

insider [ɪn'saɪdə] n.内部的人，知道内情的人

account [ə'kaʊnt] n.账号

encrypt [ɪn'krɪpt] v.加密，将……译成密码

maintain [meɪn'teɪn] vt.维持，维护

decoy ['di:kɔɪ] vt.诱骗

surveillance [sɜ:'veɪləns] n.监视，监督

legitimate [lɪ'dʒɪtɪmət] adj.合法的，合理的

v.合法

exploitation [ˌeksplɔɪ'teɪʃn] n.开发

distract [dɪ'strækt] v.转移

Honeynet ['hʌnɪnet] n.蜜网

vulnerability [ˌvʌlnərə'bɪlətɪ] n.弱点，攻击

recipe ['resəpɪ] n.处方

safeguard ['seɪfgɑ:d] vt.维护，保护，捍卫

n.安全装置，安全措施

intrusion [ɪn'tru:ʒn] n.闯入，侵扰

broadband ['brɔ:dbænd] n.宽带

nosy ['nəʊzɪ] adj.好管闲事的，爱追问的

n.好管闲事的人

widespread ['waɪdspred] adj.分布广泛的，普遍的

device [dɪ'vaɪs] n.装置，设备

properly ['prɒpəlɪ] adv.适当地，完全地

unfortunately [ʌn'fɔ:tʃənətlɪ] adv.不幸地

passcode [pɑ:skəʊd] n.密码

illicit [ɪ'lɪsɪt] adj.违法的

spyware ['spaɪweə] n.间谍软件

habit ['hæbɪt] n.习惯，习性

dangerous ['deɪndʒərəs] adj.危险的

capture	[ˈkæptʃə]	*vt.*捕获
substantial	[səbˈstænʃl]	*adj.*实质的，真实的，充实的
notoriously	[nəʊˈtɔ:rɪəslɪ]	*adv.*声名狼藉地，臭名昭著地，众所周知地
remove	[rɪˈmu:v]	*vt.*删除，移去
reputable	[ˈrepjətəbl]	*adj.*声誉好的，有声望的，有好评的
stalker	[ˈstɔ:kə]	*n.* 跟踪者
identity	[aɪˈdentətɪ]	*n.*身份
activity	[ækˈtɪvətɪ]	*n.*行动，行为
anonymous	[əˈnɒnɪməs]	*adj.*匿名的
achievable	[əˈtʃi:vəbl]	*adj.*做得成的，可完成的

Phrases

consist of	由……组成
computer network	计算机网络
a variety of	多种的
government agency	政府机构
start with …	以……开始
one-factor authentication	单因素认证，单身份验证
two-factor authentication	双因素认证，双重身份验证
three-factor authentication	三因素认证，三重身份验证
retinal scan	虹膜扫描
fail to…	未能……
computer worm	计算机蠕虫
antivirus software	防病毒软件
early warning	预警
keep an eye on	密切注视，照看
protect from	保护
network component	网络元件
away from	远离
insist on	坚持
keylogger software	键盘记录软件
WiFi hotspot	WiFi 热点
social network	社交网络
proxy server	代理服务器

Abbreviations

ID (IDentification, IDentity)	身份，标识符
IPS (Intrusion Prevention System)	入侵防御系统
WiFi (Wireless Fidelity)	基于 IEEE 802.11b 标准的一种无线局域网

IP (Internet Protocol) 网际协议

Notes

[1] Network security involves the authorization of access to data in a network, which is controlled by the network administrator.

本句中，which is controlled by the network administrator 是一个非限定性定语从句，修饰 the authorization of access to data in a network，对其进行补充说明。

[2] Though effective to prevent unauthorized access, this component may fail to check potentially harmful content, such as computer worms or Trojan horses being transmitted over the network.

本句中，Though effective to prevent unauthorized access 是一个让步状语，可以将其扩展为一个让步状语从句 Though this component is effective to prevent unauthorized access。such as computer worms or Trojan horses 是对 potentially harmful content 的举例说明。being transmitted over the network 作定语，修饰和限定 computer worms or Trojans，可以将其扩展为一个定语从句 which are transmitted over the network。

[3] Newer systems combining unsupervised machine learning with full network traffic analysis can detect active network attackers from malicious insiders or targeted external attackers that have compromised a user machine or account.

本句中，combining unsupervised machine learning with full network traffic analysis 是一个现在分词短语作定语，修饰和限定 Newer systems。可以将其扩展为一个定语从句 which combines unsupervised machine learning with full network traffic analysis。that have compromised a user machine or account 是一个定语从句，修饰和限定 targeted external attackers。

[4] Techniques used by the attackers that attempt to compromise these decoy resources are studied during and after an attack to keep an eye on new exploitation techniques.

本句中，Techniques 是主语，are studied 是谓语。used by the attackers that attempt to compromise these decoy resources 是一个过去分词短语作定语，修饰和限定 Techniques。在该短语中，that attempt to compromise these decoy resources 是一个定语从句，修饰和限定 the attackers。during and after an attack 是时间状语，to keep an eye on new exploitation techniques 是目的状语，都修饰谓语 are studied。

[5] All spyware on a computer attempts to function without the knowledge of people using it, thereby posing a substantial security risk.

本句中，using it 是一个现在分词短语作定语，修饰和限定 people。可以将其扩展为一个定语从句 who use it。thereby posing a substantial security risk 作结果状语。

Text B

Text B
Firewall

Firewall

A firewall is a part of a computer system or network that is designed to block unauthorized access while permitting authorized communications. It is a device or set of devices which is configured to permit or deny computer applications based upon a set of rules and other criteria (Figure 8-1).

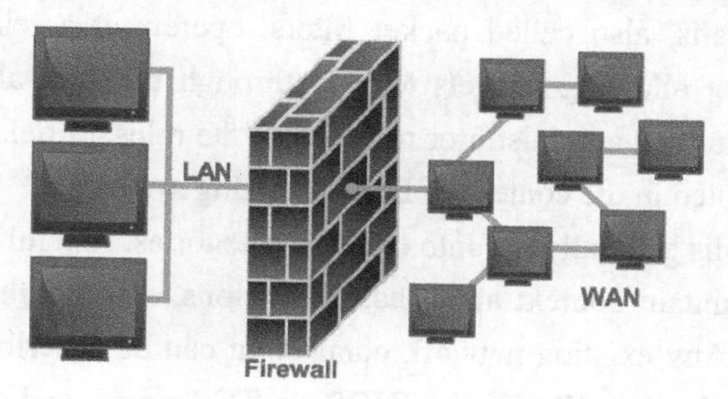

Figure 8-1 An illustration of where a firewall would be located in a network

Firewalls can be implemented in either hardware or software, or a combination of both. Firewalls are frequently used to prevent unauthorized Internet users from accessing private networks connected to the Internet, especially Intranets. All messages entering or leaving the Intranet pass through the firewall, which examines each message and blocks those that do not meet the specified security criteria.

There are several types of firewall techniques.

(1) Packet filter: packet filtering inspects each packet passing through the network and accepts or rejects it based on user-defined rules. Although difficult to configure, it is fairly effective and mostly transparent to its users. It is susceptible to IP spoofing.

(2) Application gateway: applies security mechanisms to specific applications, such as FTP and Telnet servers. This is very effective, but can impose performance degradation.

(3) Circuit-level gateway: applies security mechanisms when a TCP or UDP connection is established. Once the connection has been made, packets can flow between the hosts without further checking.

(4) Proxy server: intercepts all messages entering and leaving the network. The proxy server effectively hides the true network addresses.

1. Function

A firewall is a dedicated appliance, or software running on a computer, which inspects network traffic passing through it, and denies or permits passage based on a set of rules/criteria.

It is normally placed between a protected network and an unprotected network and acts like a gate to protect assets to ensure that nothing private goes out and nothing malicious comes in.

A firewall's basic task is to regulate some of the flow of traffic between computer networks of different trust levels. Typical examples are the Internet which is a zone with no trust and an Internal network which is a zone of higher trust. A zone with an intermediate trust level, situated between the Internet and a trusted internal network, is often referred to as a "perimeter network" or Demilitarized zone (DMZ).

A firewall's function within a network is similar to physical firewalls with fire doors in building construction. In the network, it is used to prevent network intrusion to the private network. In the building, it is intended to contain and delay structural fire from spreading to adjacent structures.

2. Types

There are several classifications of firewalls depending on where the communication is taking place, where the communication is intercepted and the state that is being traced.

2.1 Network layer

Network layer firewalls, also called packet filters, operate at a relatively low level of the TCP/IP protocol stack, not allowing packets to pass through the firewall unless they match the established rule set. The firewall administrator may define the rules, or default rules may apply. The term "packet filter" originated in the context of BSD operating systems.

Network layer firewalls generally fall into two sub-categories, stateful and stateless.

Stateful firewalls maintain context about active sessions, and use that "state information" to speed packet processing. Any existing network connection can be described by several properties, including source and destination IP address, UDP or TCP ports, and the current stage of the connection's lifetime (including session initiation, handshaking, data transfer, or completion connection). If a packet does not match an existing connection, it will be evaluated according to the rule set for new connections. If a packet matches an existing connection based on comparison with the firewall's state table, it will be allowed to pass without further processing.

Stateless firewalls require less memory, and can be faster for simple filters that require less time to filter than to look up a session. They may also be necessary for filtering stateless network protocols that have no concept of a session. However, they cannot make more complex decisions based on what stage communications between hosts have reached.

Modern firewalls can filter traffic based on many packet attributes like source IP address, source port, destination IP address or port, destination service like WWW or FTP. They can filter based on protocols, TTL values, netblock of originator, of the source, and many other attributes.

2.2 Application-layer

Application-layer firewalls work on the application level of the TCP/IP stack (i.e., all browser traffic, or all telnet or FTP traffic), and may intercept all packets traveling to or from an application. They block other packets (usually dropping them without acknowledgment to the sender). In principle, application layer firewalls can prevent all unwanted outside traffic from reaching protected machines.

On inspecting all packets for improper content, firewalls can restrict or prevent outright the spread of networked computer worms and Trojan horses. The additional inspection criteria can add extra latency to the forwarding of packets to their destination.

2.3 Proxies

A proxy device (running either on dedicated hardware or as software on a general-purpose machine) may act as a firewall by responding to input packets (connection requests, for example) in the manner of an application, whilst blocking other packets.

Proxies make tampering with an internal system from the external network more difficult, and misuse of one internal system would not necessarily cause a security breach exploitable from outside the firewall (as long as the application proxy remains intact and properly configured). Conversely, intruders may hijack a publicly-reachable system and use it as a proxy for their own purposes; the proxy then masquerades as that system and connects to other internal machines. While use of internal address spaces enhances security, crackers may still employ methods such as IP spoofing to attempt to pass packets to a target network.

2.4 Network address translation

Firewalls often have network address translation (NAT) functionality, and the hosts protected behind a firewall commonly have addresses in the "private address range", as defined in RFC 1918. Firewalls often have such functionality to hide the true address of protected hosts. Originally, the NAT function was developed to address the limited number of IPv4 routable addresses that could be used or assigned to companies or individuals, as well as to reduce both the amount and cost of obtaining enough public addresses for every computer in an organization. Hiding the addresses of protected devices has become an increasingly important defense against network reconnaissance.

New Words

deny	[dɪ'naɪ]	*v*.否认，拒绝
criteria	[kraɪ'tɪərɪə]	*n*.标准
combination	[ˌkɒmbɪ'neɪʃn]	*n*.结合，联合，合并
inspect	[ɪn'spekt]	*v*.检查
reject	[rɪ'dʒekt]	*vt*.拒绝，抵制，否决，丢弃
transparent	[træns'pærənt]	*adj*.透明的，显然的，明晰的
spoof	[spu:f]	*v*.哄骗，欺骗
degradation	[ˌdegrə'deɪʃn]	*n*.降级，降格，退化
intercept	[ˌɪntə'sept]	*vt*.中途阻止，截取
hide	[haɪd]	*v*.隐藏，掩藏，隐瞒
intermediate	[ˌɪntə'mi:dɪət]	*adj*. 中间的 *n*.媒介
situated	['sɪtjʊeɪtɪd]	*adj*.位于，被置于……境遇
originate	[ə'rɪdʒɪneɪt]	*vi*.起源，发生
stateless	['steɪtləs]	*a*.无状态的
property	['prɒpətɪ]	*n*.性质，特性
stage	[steɪdʒ]	*n*.发展的进程、阶段或时期
handshake	['hændʃeɪk]	*n*.握手
acknowledgment	[ək'nɒlɪdʒmənt]	*n*.承认
sender	['sendə]	*n*.寄件人，发送人
principle	['prɪnsəpl]	*n*.法则，原则，原理
worm	[wɜ:m]	*n*.蠕虫
latency	['leɪtənsɪ]	*n*.反应时间；潜伏，潜在，潜伏物
tamper	['tæmpə]	*vi*.篡改
misuse	[ˌmɪs'ju:z]	*v*.&*n*.误用，错用，滥用
breach	[bri:tʃ]	*n*.破坏，裂口 *vt*.打破，突破
exploitable	[ɪks'plɔɪtəbl]	*adj*.可开发的，可利用的
intact	[ɪn'tækt]	*adj*.完整无缺的

hijack	['haɪdʒæk]	*vt.*劫持
masquerade	[ˌmæskə'reɪd]	*v.*化装；伪装
cracker	['krækə]	*n.*骇客
increasingly	[ɪn'kri:sɪŋ lɪ]	*adv.*日益，愈加
reconnaissance	[rɪ'kɒnɪsəns]	*n.*侦察，搜索

Phrases

base upon	根据，依据
pass through	经过，通过
packet filter	包过滤
be susceptible to …	对……敏感，可被……
circuit-level gateway	电路层网关
perimeter network	外围网络，非军事区网络
fire door	防火门
spread to	传到，波及，蔓延到
stateless firewall	无状态防火墙
take place	发生，进行
rule set	规则集
fall into	分成，属于
active session	有效对话期间，工作时间
state table	状态表
routable address	可路由地址

Abbreviations

DMZ (DeMilitarized Zone)	非军事区
BSD (Berkeley Software Distribution)	伯克利软件发布
TTL (Time To Live)	存活时间
NAT (Network Address Translation)	网络地址转换

Word Building

常用前缀列举如下：

1. semi-表示：半

conductor 导体 —— semiconductor 半导体

automatic 自动的 —— semiautomatic 半自动的

diameter 直径 —— semidiameter 半径

2. sub-表示：在……底下；亚，次，分

directory 目录 —— subdirectory 子目录

way 路，道 —— subway 地道，地铁

head 标题 —— subhead 副标题，小标题
area 区域 —— subarea 分区
3. super-表示：超
market 市场 —— supermarket 超级市场
power 功率 —— superpower 超功率
profit 利润 —— superprofit 超额利润
highway 高速公路 —— superhighway 超级高速公路
4. tele-表示：远、电
vision 视力 —— television 电视
graph 曲线图，图表 —— telegraph 电报
meter 仪表 —— telemeter 遥测计
5. trans-表示：转换；横过；超
national 国家的 —— transnational 跨国的，超越国界的
plant 种植 —— transplant 移植
form 形式 —— transform 转化，改变
personal 个人的 —— transpersonal 非个人的，超越个人的
6. ultra-表示：超过、极端
short 短的 —— ultrashort 超短（波）的
red 红的 —— ultrared 红外线的
speed 速度 —— ultraspeed 超高速的
microscope 显微镜 —— ultramicroscope 超微显微镜
7. un-表示：反、不、非
format 格式化 —— unformat 未格式化
delete 删除 —— undelete 不删除，恢复
install 安装 —— uninstall 拆除，卸去
important 重要的 —— unimportant 不重要的
8. under-表示：在……下；次于，低于；不足
ground 地，地面 —— underground 地下的
write 写 —— underwrite 写于……之下
agent 代理人 —— underagent 副代理人
size 尺寸，大小 —— undersize 不够大的，小于一般尺寸的
9. vice-表示：副的
chairman 主席 —— vice-chairman 副主席
manager 经理 —— vice-manager 副经理

Career Training

邀 请 函

邀请函是一种重要的社交书信，包括正式和非正式两种。正式的邀请函有固定的格式，

一般用第三人称书写；非正式的邀请函格式不严格，使用第一人称或第二人称。中英文邀请函各自均有固定的格式和措辞，不能简单地逐句翻译。正式英文邀请函从头至尾都采用第三人称，译成中文时，一般应改用第一人称；英文邀请函中星期应写在日期之前，译成中文时，星期应写在日期后面的括号内。

以会议邀请卡（conference invitation card）为例，卡中应清楚注明会议的地点与时间，以及应该在什么时间回函。邀请卡中的 RSVP 为法语“Répondez s’il vous plaît” 的缩写，即 “Please reply”（请回函）。

请看下面会议邀请卡范例：

The honor of your presence
is requested
At National Conference on Multi-media Technology
3:00 p.m. Saturday the tenth of March 2020
Hilton Hotel
188 Xinhua Road
Shanghai, China
RSVP by March 1st on enclosed card

会议邀请卡的内容翻译如下：

“全国多媒体技术研讨会”定于 2020 年三月十日（星期六）下午三点在上海市新华路 188 号希尔顿酒店举行，敬请参加。请在三月一日前回函。

软件水平考试试题解析

【真题再现】

从供选择的答案中选出应填入下列英语文句中____内的正确答案，把编号写在答卷的对应栏内。

A database system gives us a way of __A__ together specific pieces or lists of __B__ that are relevant to us in our jobs or our lives. It also provides a way to __C__ and maintain that information in a central place. The first commercial computers were really __D__ more than dedicated database machine used to gather, sort and report on census information. To this day, one of the most common reasons for purchasing a computer is to __E__ a database system.

供选择的答案

A：①gather　②gathering　③get　④getting

B：①data　②information　③mail　④message

C：①build　②copy　③remember　④store

D：①anything　②nothing　③something　④thing

E：①find　②load　③run　④install

【答案】A：②　B：②　C：④　D：②　E：③

【试题解析】

A：gather together 的意思是“集合，集聚，收集”，如：gather together one’s belongings 的

意思是“收集好某人的东西”。gather together information，即“收集信息”。get together 的意思是“相聚，聚会”，如“When can we get together?”，意为“我们什么时候聚会？”。根据句子结构，介词 of 后面应该用动名词形式，再根据句子的意思，应选②。

B：data 的意思是“数据”，information 的意思是“信息”，mail 的意思是“邮件”，message 的意思是“消息，口信，讯息”。根据句子的意思，与我们工作和生活有关的应该是信息。故选②。

C：build 的意思是“建立”，copy 的意思是“复制，拷贝”，remember 的意思是“记住，牢记”，store 的意思是“存储”。该句的意思是“数据库系统也提供了存储和维护信息的方式”。故选④。

D：nothing more than 的意思是“仅仅，只不过”。故选②。

E：find 的意思是“查找”，load 的意思是“装载”，run 的意思是“运行”，install 的意思是“安装”。根据句意，购买计算机的一个最普遍的原因是要运行数据库系统，应选③。

【参考译文】

数据库系统提供了一种可以把与工作和生活相关的信息细节或片段聚集起来的方式，还提供了在中心机构存储和维护信息的方式。第一批商用计算机只不过是专用的数据库机器，用来收集、排序、报告人口普查信息。时至今日，购买计算机的一个最普遍的原因是要运行数据库系统。

Exercises

[Ex. 1] 根据 Text A 回答以下问题。

1) What does network security consist of?

2) What does network security involve?

3) What is the most common and simple way of protecting a network resource?

4) What does a honeypot encourage attackers to do?

5) What is the purpose of a honeynet?

6) What does the most basic but often overlooked element of network security involve?

7) What are the few common-sense best practices in password management which will greatly improve the security protection on a computer network, mentioned in the passage?

8) What is one of the most dangerous forms of spyware mentioned in the passage? What does it do?

9) Why do security experts recommend installing and running reputable anti-spyware software

on computer networks?

10) How can much about a person's identity can be tracked online even in the virtual world?

[Ex. 2] 根据 Text B 回答以下问题。

1) What is a firewall?

2) What can firewalls be implemented in?

3) What are the types of firewall techniques mentioned in the passage?

4) What is a firewall's basic function?

5) What is a firewall's function within a network similar to?

6) What do network layer firewalls do? What did the term "packet filter" originate?

7) What are the two sub-categories network layer firewalls generally fall into? What do they do respectively?

8) Where do application-layer firewalls work? What may they do?

9) What may a proxy device (running either on dedicated hardware or as software on a general-purpose machine) act as?

10) What was the NAT function developed to do originally?

[Ex. 3] 把下列句子翻译为中文。

We maintain strict security standards and procedures to prevent unauthorized access to information about you.

No one can access your data without the password.

At a high level, we apply authentication and authorization rigorously.

You should back up your data often.

You need to have system administrator privileges to do this.

Do not disclose your password to anyone.

This increases the communication between the database and the application.

If you enter an incorrect username or password, you should see a login error message.

The bus in this computer can transmit data between any two components of the system.

You can encrypt the entire message, so that nobody but the recipient can read it.

[Ex. 4] 软件水平考试真题自测。

从供选择的答案中选出应填入下列英语文句中____内的正确答案，把编号写在答卷的对应栏内。

Because Web servers are platform and application ____A____, they can send or request data from legacy or external applications including databases. All replies, once converted into ____B____ mark-up language, can then be transmitted to a ____C____. Used in this way, intranets can ____D____ lower desktop support costs, easy links with legacy applications and databases and, ____E____ all, ease of use.

供选择的答案：

A：①coupled ②dependent ③independent ④related

B：①ciphertext ②hypertext ③plaintext ④supertext

C：①browser ②repeater ③router ④server

D：①off ②offer ③office ④officer

E：①abort ②about ③above ④around

[Ex. 5] 听短文填空。

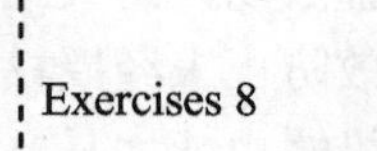

A password is a word or string of characters used for user authentication to prove ____1____ or access approval to gain access to a resource, which is to be kept secret from those not allowed access.

The use of passwords is known to be ancient. Sentries would challenge those wishing to enter an area or approaching it to supply a ____2____ or watchword, and would only allow a person or group to pass if they knew the password. In modern times, user names and passwords are commonly used by people during a log in process that ____3____ access to protected computer operating systems, mobile phones, cable TV decoders, automated teller machines (ATMs), etc. A ____4____ computer user has passwords for many purposes: logging into accounts, retrieving e-mail, accessing applications, ____5____, networks, Web sites, and even reading the morning newspaper online.

Despite the name, there is no need for passwords to be actual words; indeed passwords which are not actual words may be ____6____ to guess, a desirable property. Some passwords are formed from multiple words and may more accurately be called a passphrase. The terms passcode and passkey are sometimes used when the ____7____ information is purely numeric, such as the personal identification number (PIN) commonly used for ATM access. Passwords are generally short enough to be easily memorized and typed.

Most organizations ____8____ a password policy that sets requirements for the composition and usage of passwords, typically dictating minimum length, required categories (e.g. upper and lower case, numbers, and special characters), prohibited ____9____ (e.g. own name, date of birth, address, telephone number). Some governments have national authentication frameworks that ____10____ requirements for user authentication to government services, including requirements for passwords.

Reference Translation

网 络 安 全

网络安全包括为防止和监控未经授权的访问、滥用、修改或拒绝计算机网络和可访问网络资源而采取的政策和做法。网络安全涉及访问网络中数据的授权，这由网络管理员控制。用户选择或被分配一个 ID、密码或其他认证信息，让他们能在其权限范围内访问信息和程序。网络安全涉及在日常工作中使用的各种公共和私人计算机网络，也涉及企业、政府机构和个人之间的业务和通信。网络可以是私有的（例如公司内的网络），也可以是对公众开放。网络安全涉及组织、企业和其他类型的机构。如其名称所示，网络安全保护网络，保护和监控正在进行的操作。保护网络资源的最常见和简单的方法是为网络资源分配唯一的名称和相应的密码。

网络安全始于认证，通常使用用户名和密码。仅需要一个详细的用户名的认证，有时被称为单因素认证；也能使用双因素认证，也就是使用用户拥有的东西（例如，安全令牌、ATM 卡或移动电话）；还可使用三因素认证，用户使用某些东西（例如，指纹或虹膜扫描）证明他们是谁。

一旦认证成功，防火墙就执行访问策略，例如，允许网络用户访问什么服务。虽然访问策略可以有效防止未经授权的访问，但此组件可能无法检查可能有害的内容（例如通过网络传输的计算机蠕虫或木马）。防病毒软件或入侵防御系统（IPS）有助于检测和禁止此类恶意软件的操作。入侵检测系统还可以监视网络（如 wireshark 流量），并且可以记录下来用于以后的审计和高级分析。将无监督机器学习与全网络流量分析相结合的新系统，可以检测主动的网络攻击者，这些攻击者来自内部或外部，目的在于恶意破坏用户机器或账户。

可以加密两个网络主机之间的通信以保持隐私。

蜜罐，其本质是作为诱饵的网络可访问资源，可以被部署在网络中作为监视和预警工具，因此蜜罐通常不用于合法目的的访问。通过在攻击期间和攻击后研究攻击者所使用的尝试破坏这些诱饵资源的技术，可以了解新的攻击技术。这样的研究可用于进一步提高被蜜罐保护的实际网络的安全性。蜜罐还可以将攻击者的注意力从合法的服务器引开，鼓励攻击者将时间和精力花费在诱饵服务器上，同时分散他们对真实服务器上的数据的注意力。类似于蜜罐，蜜网也是故意设置了漏洞的网络。其目的也是吸引攻击，以便研究攻击者的方法，并且可以使用研究结果来提高网络安全性。蜜网通常包含一个或多个蜜罐。

由于所有重要的个人和商业数据每天都在计算机网络上共享，因而安全对网络而言至为重要。没有哪个方法可以完全保护网络使之不被入侵。随着攻击和防御的方法越来越复杂，网络安全技术随着时间的推移而不断改进和发展。

1. 物理网络安全

网络安全最基本但经常被忽视的因素是保护硬件免受盗窃或物理入侵。

公司花费大量资金将其网络服务器、网络交换机和其他核心网络部件保护在安全的设施中。虽然这些措施对于家庭用户来说不切实际，但仍应将其宽带路由器保留在私人地点，远离爱管闲事的邻居和客人。

移动设备的广泛使用使得物理安全变得更加重要。小物件特别容易被遗忘在旅行站或从口袋里掉出来。

最后，当把移动设备借给别人时，让其在视线之内：恶意的人可以偷窃个人数据、安装监控软件或者在几分钟的无人注意的时间内“黑入”手机。

2. 密码保护

如果应用得当，密码能够非常有效地提高网络的安全性。不幸的是，一些人不认真管理密码，并坚持在他们的系统和网络上使用不良的、弱的（就是容易猜到的）密码，如“123456”。

以下方法虽然只是常用的密码管理方法，但却可以大大提高计算机网络的安全性：

- 在加入网络的所有设备上设置强密码或一次性密码。
- 更改网络路由器的默认管理员密码。
- 不要经常与他人共享密码；如果可能，为朋友和来访者设置访客的网络访问密码。当知道这些密码的人较多时，更改密码。

3. 间谍软件

即使没有物理访问设备或完全不知道任何网络密码，被称为间谍软件的非法程序也可以感染计算机和网络，通常访问网站时就可能会感染。互联网上存在大量间谍软件。某些间谍软件监视个人的计算机使用和网络浏览习惯，并将这些信息报告给广告公司，以便这些公司使用它们来创建更有针对性的广告。一些间谍软件尝试窃取个人数据。其中最危险的间谍软件形式之一是键盘监控软件，它捕获并发送一个人所有键盘按键的历史记录，也是捕获密码和信用卡号码的理想工具。计算机上所有的间谍软件都试图在用户不知情的情况下工作，从而造成相当大的安全风险。

由于间谍软件非常难以检测和删除，因而网络安全专家建议在计算机网络上安装和运行著名的反间谍软件程序。

4. 在线隐私

个人追踪者、身份信息窃取者甚至政府机构，都在监控人们的网络习惯和行动，这远远超出了基本的间谍软件的范围。例如，使用来自通勤车和汽车的 WiFi 热点，能够获取个人的位置。即使在虚拟世界中，通过网络的 IP 地址和社交网络活动，也可以在线跟踪很多个人身份信息。

保护个人在线隐私的技术包括匿名 Web 代理服务器，尽管当今的技术不能完全保护在线隐私。

Unit 9

Text A

Cloud Computing Glossary

CDN (content delivery network)

A system consisting of multiple computers that contain copies of data, which are located in different places on the network so clients can access the copy closest to them.[1]

Cloud

A metaphor for a global network, first used in reference to the telephone network and now commonly used to represent the Internet.

Cloud application

A software application that is never installed on a local machine — it's always accessed over the Internet. The "top" layer of the cloud pyramid is where "applications" are run and interacted with via a Web-browser. Cloud applications are tightly controlled, leaving little room for modification. Examples include Gmail or SalesForce.com.

Cloud as a service (CaaS)

A cloud computing service that has been opened up into a platform that others can build upon.

Cloud bridge

Running an application in such a way that its components are integrated within multiple cloud environments (which could be any combination of internal/private and external/public clouds).

Cloud broker

An entity that creates and maintains relationships with multiple cloud service providers. It acts as a liaison between cloud services customers and cloud service providers, selecting the best provider for each customer and monitoring the services.

Cloudburst

What happens when your cloud has an outage or security breach and your data is unavailable.

Cloudcenter

A datacenter in the "cloud" utilizing standards-based virtualized components as a datacenter-like infrastructure; example: a large company, such as Amazon, that rents its infrastructure.

Cloud computing

A computing capability that provides an abstraction between the computing resource and its underlying technical architecture (e.g., servers, storage, networks), enabling convenient, on-demand network access to a shared pool of configurable computing resources that can be rapidly provisioned and released with minimal management effort or service provider interaction.[2] This definition states that clouds have five essential characteristics: on-demand self-service, broad network access, resource pooling, rapid elasticity and measured service. Narrowly speaking, cloud computing is client-server computing that abstract the details of the server away; one requests a service (resource), not a specific server (machine). Cloud computing enables infrastructure as a service (IaaS), platform as a service (PaaS), and software as a service (SaaS). Cloud computing means that infrastructure, applications, and business processes can be delivered to you as a service over the Internet (or your own network).

Cloud computing security

Cloud computing security is the set of control-based technologies and policies designed to protect information, data applications and infrastructure associated with cloud computing use.

Because of the cloud's very nature as a shared resource, identity management, privacy and access control are of particular concern. With more organizations using cloud computing and associated cloud providers for data operations, proper security in these and other potentially vulnerable areas have become a priority for organizations contracting with a cloud computing provider.

Cloud computing security processes should address the security controls the cloud provider will incorporate to maintain the customer's data security, privacy and compliance with necessary regulations.[3] The processes will also likely include a business continuity and data backup plan in the case of a cloud security breach.

Cloud hosting

A type of Internet hosting where the client leases virtualized, dynamically scalable infrastructure on an as-needed basis. Users frequently have the choice of operating system and other infrastructure components. Typically cloud hosting is self-service, billed hourly or monthly, and controlled via a Web interface or API.

Cloud infrastructure

The "bottom" layer—or foundation—of the cloud pyramid is the delivery of computer infrastructure through paravirtualization. This includes servers, networks and other hardware appliances delivered as either infrastructure Web services or "cloudcenters". Full control of the infrastructure is provided at this level. Examples include GoGrid or Amazon Web services.

Cloud operating system

A computer operating system that is specially designed to run in a provider's datacenter and be delivered to the user over the Internet or another network. Windows Azure is an example of a cloud operating system. The term is also sometimes used to refer to cloud-based client operating systems such as Google's Chrome OS.

Cloud-oriented architecture (COA)

A cloud-oriented architecture (COA) is a conceptual model encompassing all elements in a cloud environment. In information technology, architecture refers to the overall structure of an information system and the interrelationships of entities that make up that system.

A cloud-oriented architecture is related to both service-oriented architectures (SOA) and event-driven architectures (EDA) and is a combination of two other architectural models: the resource-oriented architecture (ROA) and the hypermedia-oriented architecture (HOA). A ROA is based on the idea that any entity that can be assigned a uniform resource identifier (URI) is a resource. As such, resources include not only infrastructure elements such as servers, computers and other devices, but also Web pages, scripts and JSP/ASP pages, and other entities. Hypermedia extends the notion of the hypertext link to include links among any set of multimedia objects, including sound, video, and virtual reality.

Cloud platform

The "middle" layer of the cloud pyramid which provides a computing platform or framework (e.g., .NET, Ruby on Rails, or Python) as a service or stack. Control is limited to that of the platform or framework, but not at a lower level (server infrastructure). Examples include Google AppEngine or Microsoft Azure.

Cloud portability

The ability to move applications (and often their associated data) across cloud computing environments from different cloud providers, as well as across private or internal cloud and public or external clouds.

Cloud provider

A company that provides cloud-based platform, infrastructure, application, or storage services to other organizations and/or individuals, usually for a fee.

Cloud pyramid

A visual representation of cloud computing layers where differing segments are broken out by functionality. Simplified version includes infrastructure, platform and application layers.

Cloud servers

Virtualized servers running Windows or Linux operating systems that are instantiated via a Web interface or API. Cloud servers behave in the same manner as physical ones and can be controlled at an administrator or root level, depending on the server type and cloud hosting provider.

Cloud standards

A standard is an agreed-upon approach for doing something. Cloud standards ensure interoperability, so you can take tools, applications, virtual images, and more, and use them in another cloud environment without having to do any rework. Portability lets you take one application or instance running on one vendor's implementation and deploy it on another vendor's implementation.

Cloud storage

A service that allows customers to save data by transferring it over the Internet or another network to an offsite storage system maintained by a third party.

Cloudware

A general term referring to a variety of software, typically at the infrastructure level, that enables building, deploying, running or managing applications in a cloud computing environment.

Consumption-based pricing model

A pricing model whereby the service provider charges its customers based on the amount of the service the customer consumes, rather than a time-based fee. For example, a cloud storage provider might charge per gigabyte(GB) of information stored.

Customer self-service

A feature that allows customers to provision, manage, and terminate services themselves, without involving the service provider, via a Web interface or programmatic calls to service APIs.

Elasticity and scalability

The cloud is elastic, meaning that resource allocation can get bigger or smaller depending on demand. Elasticity enables scalability, which means that the cloud can scale upward for peak demand and downward for lighter demand.[4] Scalability also means that an application can scale when adding users and when application requirements change.

External cloud

Public or private cloud services that are provided by a third party outside the organization. A cloud computing environment that is external to the boundaries of the organization.

Hosted application

An Internet-based or Web-based application software program that runs on a remote server and can be accessed via an Internet-connected PC or thin client.

Hybrid cloud

A networking environment that includes multiple integrated internal and/or external providers. Hybrid clouds combine aspects of both public and private clouds.

Identity management

Managing personal identity information so that access to computer resources, applications, data and services is controlled properly.

Infrastructure as a service (IaaS)

Cloud infrastructure services or "Infrastructure as a Service"(IaaS) delivers computer infrastructure, typically a platform virtualization environment, as a service. Rather than purchasing servers, software, data center space or network equipment, clients instead buy those resources as a fully outsourced service. The service is typically billed on a utility computing basis and amount of resources consumed (and therefore the cost) will typically reflect the level of activity. It is an evolution of Web hosting and virtual private server offerings.

Internal cloud

A type of private cloud whose services are provided by an IT department to those in its own organization.

On-demand service

A model by which a customer can purchase cloud services as needed; for instance, if customers need to utilize additional servers for the duration of a project, they can do so and then drop back to the previous level after the project is completed.

Personal cloud

Synonymous with something called MiFi, a personal wireless router. It takes a mobile wireless data signal and translates it to WiFi.

Platform as a service (PaaS)

Platform as a service — cloud Platform services, whereby the computing platform (operating system and associated services) is delivered as a service over the Internet by the provider. The PaaS layer offers black box services with which developers can build applications on top of the compute infrastructure. This might include developer tools that are offered as a service to build services, or data access and database services, or billing services.

Private cloud

Virtualized cloud data centers inside your company's firewall. It may also be a private space dedicated to your company within a cloud provider's data center. An internal cloud behind the organization's firewall. The company's IT department provides software and hardware as a service to its customers — the people who work for the company. Vendors love the words"private cloud".

Public cloud

Services offered over the public Internet and available to anyone who wants to purchase the service.

SaaS (software as a service)

Cloud application services, whereby applications are delivered over the Internet by the provider, so that the applications don't have to be purchased, installed, and run on the customer's computers. SaaS providers were previously referred to as ASP (application service providers). In the SaaS layer, the service provider hosts the software so you don't need to install it, manage it, or buy hardware for it. All you have to do is connect and use it. SaaS Examples include customer relationship management as a service.

Self-service provisioning

Cloud customers can self-provision cloud services without going through a lengthy process. You request an amount of computing, storage, software, process, or more from the service provider. After you use these resources, they can be automatically deprovisioned.

Standardized interfaces

Cloud services should have standardized APIs, which provide instructions on how two application or data sources can communicate with each other.[5] A standardized interface lets the customer more easily link cloud services together.

Vertical cloud

A cloud computing environment that is optimized for use in a particular industry, such as health care or financial services.

Virtual private cloud (VPC)

The term virtual private cloud (VPC) describes a concept that is similar to, and derived from, the familiar concept of a virtual private network (VPN), but applied to cloud computing. It is the notion of turning a public cloud into a virtual private cloud.

VPC is a hybrid model of cloud computing in which a private cloud solution is provided within a public cloud provider's infrastructure.

VPC is a cloud computing service in which a public cloud provider isolates a specific portion of their public cloud infrastructure to be provisioned for private use. The VPC infrastructure is managed by a public cloud vendor; however, the resources allocated to a VPC are not shared with any other customer.

New Words

metaphor	['metəfə]	*n.*隐喻，暗喻，比喻
pyramid	['pɪrəmɪd]	*n.*金字塔；角锥体
		v.（使）成金字塔状；（使）渐增，（使）上涨
modification	[ˌmɒdɪfɪ'keɪʃn]	*n.*更改，修改，修正
host	[həʊst]	*vt.*托管
		*n.*主机
component	[kəm'pəʊnənt]	*n.*成分，部件
		*adj.*组成的，构成的
integrated	['ɪntɪgreɪtɪd]	*adj.*集成的，综合的，完整的
broker	['brəʊkə]	*n.*经纪人
liaison	[li'eɪzn]	*n.*联络
cloudburst	['klaʊdbɜ:st]	*n.*云破裂
outage	['aʊtɪdʒ]	*n.*断供，停止供应
unavailable	[ˌʌnə'veɪləbl]	*adj.*不可用的，无效的
datacenter	['deɪtəsentə]	*n.*数据中心
abstraction	[æb'strækʃn]	*n.*抽象，提取
configurable	[kən'fɪgərəbl]	*adj.*结构的，可配置的
interaction	[ˌɪntər'ækʃn]	*n.*交互作用
proper	['prɒpə]	*adj.*适当的，正确的，固有的，特有的
potentially	[pə'tenʃəlɪ]	*adv.*潜在地
vulnerable	['vʌlnərəbl]	*adj.*易受攻击的
incorporate	[ɪn'kɔ:pəreɪt]	*v.*合并 *adj.*合并的
continuity	[ˌkɒntɪ'nju:ətɪ]	*n.*连续性，连贯性
Internet	['ɪntənet]	*n.*互联网
dynamical	[daɪ'næmɪkəl]	*adj.*动态的
scalable	['skeɪləbl]	*adj.*可扩展的，可升级的
foundation	[faʊn'deɪʃn]	*n.*基础，根本，建立，创立
paravirtualization	['pærəvɜ:tʃʊəlaɪ'zeɪʃn]	*n.*准虚拟化，半虚拟化
conceptual	[kən'septʃʊəl]	*adj.*概念上的
encompass	[ɪn'kʌmpəs]	*v.*包围，环绕，包含或包括某事物
interrelationship	[ˌɪntərɪ'leɪʃnʃɪp]	*n.*相互关系，相互关联

entity	['entətɪ]	*n.*实体
hypermedia	[ˌhaɪpə'mi:dɪə]	*n.*超媒体
notion	['nəʊʃn]	*n.*概念，观念，想法
portability	[ˌpɔ:tə'bɪlətɪ]	*n.*可移植性，可携带，轻便
representation	[ˌreprɪzen'teɪʃn]	*n.*表示法，表现，代表
instantiate	[ɪns'tænʃɪeɪt]	*vt.*具体化，实例化
rework	[ˌri:'wɜ:k]	*vt.*重做，改写，重写
implementation	[ˌɪmplɪmen'teɪʃn]	*n.*执行，实现
offsite	[ɒf'saɪt]	*adj.*异地的
cloudware	[klaʊdweə]	*n.*云件
elasticity	[ˌɪlæ'stɪsətɪ]	*n.*弹性
scalability	[skeɪlə'bɪlətɪ]	*n.*可扩展性，可量测性
hybrid	['haɪbrɪd]	*adj.*混合的
dedicated	['dedɪkeɪtɪd]	*adj.*专门的，专用的
provision	[prə'vɪʒn]	*n.*供应，预备，规定

Phrases

telephone network	电话网
local machine	本地计算机
Cloud Bridge	云桥
internal cloud	内部云
external cloud	外部云
private cloud	私有云
public cloud	公有云
cloud computing	云计算
business process	业务流程
identity management	身份管理
contract with …	与……签订合同
data backup	数据备份
cloud operating system	云操作系统
cloud-based client operating system	基于云的客户端操作系统
hypertext link	超文本链接
virtual reality	虚拟真实，虚拟现实
third party	第三方
consumption-based pricing model	基于消费的定价模型
time-based fee	基于时间收费
platform virtualization environment	平台虚拟化环境
on-demand service	按需服务
the duration of …	在……期间

black box　黑箱
customer relationship management as a service　客户关系管理即服务

Abbreviations

CDN (Content Delivery Network)　内容分发网络
CaaS (Cloud as a Service)　云即服务
IaaS (Infrastructure as a Service)　基础设施即服务
PaaS (Platform as a Service)　平台即服务
SaaS (Software as a Service)　软件即服务
COA (Cloud-Oriented Architecture)　面向云的体系结构
SOA (Service-Oriented Architecture)　面向服务的体系结构，服务导向的体系结构
EDA (Event-Driven Architecture)　面向事件的体系结构，事件驱动的体系结构
ROA (Resource-Oriented Architecture)　面向资源的体系结构，资源驱动的体系结构
HOA(Hypermedia-Oriented Architecture)　面向超媒体的体系，超媒体驱动的体系结构
URI (Uniform Resource Identifier)　统一资源定位符
MiFi (Mobile Wireless Fidelity)　移动无线保真
ASP (Application Service Provider)　应用服务提供方
VPN (Virtual Private Network)　虚拟专用网络

Notes

[1] A system consisting of multiple computers that contain copies of data, which are located in different places on the network so clients can access the copy closest to them.

这是一个名词短语，是对 CDN (content delivery network)的解释。consisting of multiple computers 是一个动名词短语，修饰和限定 A system。that contain copies of data 是一个定语从句，修饰和限定 computers。which are located in different places on the network so clients can access the copy closest to them 是一个非限定性定语从句，对 copies of data 进行补充说明。在该从句中，so clients can access the copy closest to them 是一个目的状语，修饰谓语 are located。

[2] A computing capability that provides an abstraction between the computing resource and its underlying technical architecture (e.g., servers, storage, networks), enabling convenient, on-demand network access to a shared pool of configurable computing resources that can be rapidly provisioned and released with minimal management effort or service provider interaction.

在这个名词短语中，that provides an abstraction between the computing resource and its underlying technical architecture 是一个定语从句，修饰和限定 A computing capability。e.g., servers, storage, networks 是对 the computing resource and its underlying technical architecture 的举例说明。enabling convenient, on-demand network access to a shared pool of configurable computing resources that can be rapidly provisioned and released with minimal management effort or service provider interaction 是一个现在分词短语，作结果状语。that can be rapidly provisioned and released with minimal management effort or service provider interaction 是一个定语从句，修饰和限定 configurable computing resources。

[3] Cloud computing security processes should address the security controls the cloud provider will incorporate to maintain the customer's data security, privacy and compliance with necessary regulations.

本句中，the cloud provider will incorporate 是一个定语从句，修饰和限定 the security controls。to maintain the customer's data security, privacy and compliance with necessary regulations 是一个动词不定式短语，作目的状语，修饰谓语 should address。

[4] Elasticity enables scalability, which means that the cloud can scale upward for peak demand and downward for lighter demand.

本句中，which means that the cloud can scale upward for peak demand and downward for lighter demand 是一个非限定性定语从句，对 Elasticity enables scalability 这个句子进行补充说明。

[5] Cloud services should have standardized APIs, which provide instructions on how two application or data sources can communicate with each other.

本句中，which provide instructions on how two application or data sources can communicate with each other 是一个非限定性定语从句，对 standardized APIs 进行补充说明。

Text B

Text B
Cloud Storage

Cloud Storage

Cloud storage is a model of computer data storage in which the digital data is stored in logical pools. The physical storage spans multiple servers (sometimes in multiple locations), and the physical environment is typically owned and managed by a hosting company. These cloud storage providers are responsible for keeping the data available and accessible, and the physical environment protected and running. People and organizations buy or lease storage capacity from the providers to store user, organization, or application data.

Cloud storage services may be accessed through a colocated cloud computing service, a Web service application programming interface (API) or by applications that utilize the API, such as cloud desktop storage, a cloud storage gateway or Web-based content management systems.

Cloud storage is based on highly virtualized infrastructure and is like broader cloud computing in terms of accessible interfaces, near-instant elasticity and scalability, multi-tenancy, and metered resources. Cloud storage services can be utilized from an off-premises service (Amazon S3) or deployed on-premises (ViON Capacity Services).

Cloud storage typically refers to a hosted object storage service, but the term has broadened to include other types of data storage that are now available as a service, like block storage.

Object storage services like Amazon S3, Oracle Cloud Storage and Microsoft Azure Storage, object storage software like Openstack Swift, object storage systems like EMC Atmos, EMC ECS and Hitachi Content Platform, and distributed storage research projects like OceanStore and Vision Cloud are all examples of storage that can be hosted and deployed with cloud storage characteristics.

Cloud storage is:

- Made up of many distributed resources, but still acts as one, either in a federated or a cooperative storage cloud architecture.
- Highly fault tolerant through redundancy and distribution of data.
- Highly durable through the creation of versioned copies.
- Typically eventually consistent with regard to data replicas.

1. Advantages

Companies need only pay for the storage they actually use, typically an average of consumption during a month. This does not mean that cloud storage is less expensive, only that it incurs operating expenses rather than capital expenses.

Businesses using cloud storage can cut their energy consumption by up to 70%, making them a more green business. Also at the vendor level they are dealing with higher levels of energy so they will be more equipped with managing it in order to keep their own costs down as well.

Organizations can choose between off-premises and on-premises cloud storage options, or a mixture of the two options, depending on relevant decision criteria that is complementary to initial direct cost savings potential, for instance, continuity of operations (COOP), disaster recovery (DR), security, and records retention laws, regulations, and policies.

Storage availability and data protection is intrinsic to object storage architecture, so depending on the application, the additional technology, effort and cost to add availability and protection can be eliminated.

Storage maintenance tasks, such as purchasing additional storage capacity, are offloaded to the responsibility of a service provider.

Cloud storage provides users with immediate access to a broad range of resources and applications hosted in the infrastructure of another organization via a Web service interface.

Cloud storage can be used for copying virtual machine images from the cloud to on-premises locations or to import a virtual machine image from an on-premises location to the cloud image library. In addition, cloud storage can be used to move virtual machine images between user accounts or between data centers.

Cloud storage can be used as natural disaster proof backup, as normally there are 2 or 3 different backup servers located in different places around the globe.

Cloud storage can be mapped as a local drive with the WebDAV protocol. It can function as a central file server for organizations with multiple office locations.

2. Potential Concerns

2.1 Attack surface area

Outsourcing data storage increases the attack surface area.

When data has been distributed it is stored at more locations, increasing the risk of unauthorized physical access to the data. For example, in cloud based architecture, data is replicated and moved frequently so the risk of unauthorized data recovery increases dramatically. The manner that data is replicated depends on the service level a customer chooses and on the service provided. When encryption is in place it can ensure confidentiality. Crypto-shredding can be used when disposing of data (on a disk).

The number of people with access to the data who could be compromised (e.g., bribed, or coerced) increases dramatically. A single company might have a small team of administrators, network engineers, and technicians, but a cloud storage company will have many customers and thousands of servers, therefore a much larger team of technical staff with physical and electronic access to almost all of the data at the entire facility or perhaps the entire company. Decryption keys that are kept by the service user, as opposed to the service provider, limit the access to data by service provider employees. As for sharing multiple data in the cloud with multiple users, a large number of keys have to be distributed to users via secure channels for decryption. The keys have to be securely stored and managed by the users in their devices. Storing these keys requires rather expensive secure storage. To overcome that, key-aggregate cryptosystem can be used.

It increases the number of networks over which the data travels. Instead of just a local area network (LAN) or storage area network (SAN), data stored on a cloud requires a WAN (wide area network) to connect them both.

By sharing storage and networks with many other users/customers it is possible for other customers to access your data. Sometimes because of erroneous actions, faulty equipment, a bug and sometimes because of criminal intent. This risk applies to all types of storage and not only cloud storage. The risk of having data read during transmission can be mitigated through encryption technology. Encryption in transit protects data as it is being transmitted to and from the cloud service. Encryption at rest protects data that is stored at the service provider. Encrypting data in an on-premises cloud service on-ramp system can provide both kinds of encryption protection.

2.2 Supplier stability

Companies are not permanent and the services and products they provide can change. Outsourcing data storage to another company needs careful investigation and nothing is ever certain. Contracts set in stone can be worthless when a company ceases to exist or its circumstances change. Companies can:

- Go bankrupt.
- Expand and change their focus.
- Be purchased by other larger companies.
- Be purchased by a company headquartered in or move to a country that negates compliance with export restrictions and thus necessitates a move.
- Suffer an irrecoverable disaster.

2.3 Accessibility

Performance for outsourced storage is likely to be lower than local storage, depending on how much a customer is willing to spend for WAN bandwidth.

Reliability and availability depends on wide area network availability and on the level of precautions taken by the service provider. Reliability should be based on hardware as well as various algorithms used.

It's a given a multiplicity of data storage.

2.4 Other concerns

Security of stored data and data in transit may be a concern when storing sensitive data at a cloud storage provider.

Users with specific records-keeping requirements, such as public agencies that must retain electronic records according to statute, may encounter complications with using cloud computing and storage. For instance, the U.S. Department of Defense designated the Defense Information Systems Agency (DISA) to maintain a list of records management products that meet all of the records retention, personally identifiable information (PII), and security (information assurance, IA) requirements.

Cloud storage is a rich resource for both hackers and national security agencies. Because the cloud holds data from many different users and organizations, hackers see it as a very valuable target.

The legal aspect, from a regulatory compliance standpoint, is of concern when storing files domestically and especially internationally.

New Words

pool	[pu:l]	*n.*池
location	[ləʊ'keɪʃn]	*n.*位置，场所
environment	[ɪn'vaɪrənmənt]	*n.*环境，外界；工作平台
available	[ə'veɪləbl]	*adj.*可利用的；可获得的；能找到的
gateway	['geɪtweɪ]	*n.*门；入口；途径；网关
virtualized	['vɜ:tʃʊəlaɪzd]	*adj.*虚拟化的
tenancy	['tenənsɪ]	*n.*租用，租赁；租期
distributed	[dɪs'trɪbju:tɪd]	*adj.*分布式的
federate	['fedəreɪt]	*v.*（使）结成联盟
cooperative	[kəʊ'ɒpərətɪv]	*adj.*合作的；协助的；共同的
durable	['djʊərəbl]	*adj.*耐用的，耐久的；持久的；长期的 *n.*耐用品，耐久品
replica	['replɪkə]	*n.*复制品
consumption	[kən'sʌmpʃn]	*n.*消费
expensive	[ɪk'spensɪv]	*adj.*昂贵的，花钱多的
expense	[ɪk'spens]	*n.*费用；消耗 *vt.*向……收取费用
off-premise	['ɔ:f'premɪs]	*adj.*外部部署的
on-premise	['ɒn'premɪs]	*adj.*本地部署的
intrinsic	[ɪn'trɪnsɪk]	*adj.*固有的，内在的，本质的
immediate	[ɪ'mi:dɪət]	*adj.*立即的；直接的
outsourcing	['aʊtsɔ:sɪŋ]	*n.*外包，外购
dramatically	[drə'mætɪklɪ]	*adv.*显著地，剧烈地
disposal	[dɪ'spəʊzl]	*n.*（事情的）处置；清理

		*adj.*处理（或置放）废品的
reuse	[ˌri: ˈju:z]	*vt.*重用，复用
reallocation	[ˌri:ˌæləˈkeɪʃn]	*vt.*再分配
confidentiality	[ˌkɒnfɪˌdenʃiˈæləti]	*n.*机密性
crypto-shredding	[ˈkrɪptəʊ-ˈʃredɪŋ]	*n.*密码粉碎
compromise	[ˈkɒmprəmaɪz]	*vt.*违背（原则）；（尤指因行为不很明智）使陷入危险
bribe	[braɪb]	*v.*贿赂，行贿 *n.*贿赂
coerce	[kəʊˈɜ:s]	*vt.*控制，限制；威胁；逼迫
decryption	[di:ˈkrɪpʃn]	*n.*解密，译码
channel	[ˈtʃænl]	*n.*通道，渠道
erroneous	[ɪˈrəʊnɪəs]	*adj.*错误的，不正确的
stability	[stəˈbɪləti]	*n.*稳定（性）；稳固
investigation	[ɪnˌvestɪˈgeɪʃn]	*n.*调查，研究
bankrupt	[ˈbæŋkrʌpt]	*adj.*破产的，倒闭的 *n.*破产者 *vt.*使破产
headquarter	[ˈhedˈkwɔ:tə]	*vi.*设立总部 *vt.*将……的总部设在；把……放在总部里
irrecoverable	[ˌɪrɪˈkʌvərəbl]	*adj.*无可挽救的；不可弥补
precaution	[prɪˈkɔ:ʃn]	*n.*预防，防备，警惕；预防措施 *vt.*使提防
multiplicity	[ˌmʌltɪˈplɪsəti]	*n.*多样性
transit	[ˈtrænzɪt]	*vt.*传输
agency	[ˈeɪdʒənsɪ]	*n.* 代理；机构
encounter	[ɪnˈkaʊntə]	*vt.*不期而遇
complication	[ˌkɒmplɪˈkeɪʃn]	*n.*纠纷；混乱
hacker	[ˈhækə]	*n.*黑客
domestically	[dəˈmestɪklɪ]	*adv.*国内地；适合国内地
internationally	[ˌɪntəˈnæʃnəlɪ]	*adv.*国际性地，国际上地，国际地

Phrases

colocated cloud computing	同位云计算
cloud desktop	云桌面
content management system	内容管理系统
be based on...	基于……
block storage	块存储
object storage	对象存储
distributed storage	分布式存储

cooperative storage cloud	协同存储云
fault tolerant	容错
eventually consistent	最终一致性
data protection	数据保护
virtual machine image	虚拟机映像
natural disaster proof backup	防自然灾难备份
local drive	本地驱动器
storage space	存储空间
key-aggregate cryptosystem	密钥聚合密码系统
faulty equipment	故障设备
on-ramp system	入站匝道系统
set in stone	一成不变
export restriction	出口限制
sensitive data	敏感数据

Abbreviations

COOP (Continuity Of OPeration)	运营连续性
DR (Disaster Recovery)	灾难恢复
SAN (Storage Area Network)	存储域网
WAN (Wide Area Network)	广域网
DISA (Defense Information Systems Agency)	国防信息系统局
PII (Personally Identifiable Information)	个人身份信息
IA (Information Assurance)	信息保障

Word Building

加在单词后面的词缀叫作后缀。后缀一般不超过 4 个字母。后缀大多改变单词的词性，但词的基本意义一般不变。例如，work 是动词，表示“工作”，加-er 构成名词 worker，表示“工人”；加-able 构成动词 workable，表示“可加工的”。这一组词都与基本词义——“工作”有关。

常见的构成名词的后缀如下：

1. -age 表示：…….场所；费用；行为或行为的结果；状态；情况

mile 英里——mileage 英里数
pass 通过，通行证——passage 通道
post 邮寄，寄——postage 邮费
break （使）破，裂——breakage 破碎，破损
waste 浪费，未充分利用——wastage 耗损
short 短的——shortage 不足，短缺
advance 前进，进步——advantage 优势，优点

teen 十几岁——teenage 十几岁的时期

2. -ance 表示：性质，状况，行动，过程

abound 大量存在，有许多——abundance 丰富，充裕

ignore 忽视，对……不予理会——ignorance 忽视

intelligent 有才智的，聪明的——intelligence 智力

interfere 干涉，干预——interference 干扰，干涉

3. -ant，-ent 表示：……者

assist 帮助，协助——assistant 助手

participate 参与，参加——participant 参加者

agency 代理机构——agent 代理人

4. -er 表示：……者；……物；用于……的机械；.……人

teach 教，讲授——teacher 教师，导师

begin 开始——beginner 初学者

law 法律——lawyer 律师

read 阅读——reader 读者

consume 消耗，耗费，吃——consumer 消费者，顾客

bar 阻拦，禁止——barrier 栅栏，障碍物

cook 烹饪，煮——cooker 厨具

wash 洗，洗涤——washer 洗衣机

London 伦敦——Londoner 伦敦人

5. -ese 表示：……的人

China 中国——Chinese 中国人

Japan 日本——Japanese 日本人

6. -ess 表示：女性

actor 演员——actress 女演员

host 主人，东道主——hostess 女主人

waiter 服务员——waitress 女服务员

prince 王子——princess 公主

7. -ing 表示：动作；动作的结果；与某一动作有关者

engine 发动机，引擎——engineering 工程，工程学

feel 觉得，感到——feeling 感觉

greet 和（某人）打招呼，欢迎——greeting 问候

fish 鱼——fishing 钓鱼

8. -ism 表示：……主义；宗教；行为；……学；……术；……论；……法；……学派；具有某种特性；情况；状态

capital 资本，资金——capitalism 资本主义

hero 英雄——heroism 英雄行为

magnet 磁铁，磁石——magnetism 磁力学

hypnotic 安眠的，有催眠作用的——hypnotism 催眠术

atom 原子——atomism 原子论

cube 立方体——cubism (艺术上的)立体派

9. -ist 表示：某种主义者或某种信仰者；从事某种职业或研究的人

social 社会的，社会上的——socialist 社会主义者

type 打字——typist 打字员

science 科学——scientist 科学家

Career Training

道　歉

“人非圣贤，孰能无过。”在日常生活中，人们总会把各种各样的小错误，如迟到、打扰了别人等，因此向别人道歉也是人们生活中很常见的事情。一般来说，在英美等国，简单的道歉只需要说句“I am sorry.”就可以了。当情况稍微严重时，往往也会说“I am awfully sorry.”“I do apologize.”“Please forgive me.”或“Pardon me.”等，并且有时候致歉时还要说明原因，表示自己的行为并不是故意的，以此表示自己的诚意，从而得到对方的原谅。

道歉时的常用语如下：

I am so sorry.

I am terribly/awfully sorry about that.

I apologize.

Please forgive me.

I hope you will excuse me.

I owe you an apology.

Please accept my sincere apology.

I can’t tell you how sorry I am.

Words cannot describe how sorry I am.

I’m sorry for causing you so much inconvenience.

Sorry about the inconvenience.

I’m sorry for what I’ve done.

I honestly didn’t mean it.

I didn’t mean to do that.

I didn’t mean it that way.

It’s all my fault. I’ll try to make it up to you.

当然，一方诚心道歉后，一般来说另一方也应该表现出宽大的胸襟，友善地给予对方反应。因此一方致歉后，另一方的回答如下：

That’s all right.

No problem. We all make mistakes.

Never mind.

Forget it.

It's not your fault.
It really doesn't matter at all.
Don't worry about it.
Think no more of it.
Don't give it another thought.
Okay. I accept your apology.
That's OK.
It's all right.
It doesn't matter.
It's nothing.

软件水平考试试题解析

【真题再现】

从供选择的答案中选出应填入下列英语文句中____内的正确答案，把编号写在答卷的对应栏内。

Toolboxes and menus in many application programs were __A__ for working with the mouse. The mouse controls a pointer on the screen. You move the pointer by __B__ the mouse over a flat surface in the direction you want the pointer to move. If you run out of __C__ to move the mouse, lift it up and put it down again. The pointer moves only when the mouse is __D__ the flat surface. Moving the mouse pointer across the screen does not affect the document, the pointer simply __E__ a location on the screen. When you press the mouse button, something happens at the location of the pointer.

供选择的答案：

A：①assigned ②designed ③desired ④expressed

B：①putting ②sliding ③serving ④taking

C：①board ②place ③room ④table

D：①getting ②going ③teaching ④touching

E：①constructs ②indicates ③instructs ④processes

【答案】A：② B：② C：③ D：④ E：②

【试题解析】

A：assign 的意思是“分配，指派”，design 的意思是“设计”，desire 的意思是“渴望；欲望”。工具栏和菜单是设计过的，故选②。

B：put 的意思是“放”，slide 的意思是“滑动”，serve 的意思是“服务”，take 的意思是“拿；握住；抓住”。移动指针的方式是操作鼠标，操作鼠标的正确方法是在平面上滑动鼠标，故选②。

C：run out of 的意思是“用完，耗尽”，run out of room 是指地方不够用，故选③。

D：根据句意，只有鼠标器接触到一个平面时鼠标才能移动，其他词不合题意，故选④。

E：construct 的意思是“构成；建造”，indicate 的意思是“指出，显示”，instruct 的意思是“指示，命令”，process 的意思是“加工，处理”。因为鼠标指针仅起到屏幕定位的作用，所以选②。

【参考译文】

在许多应用程序中，设计工具栏和菜单是为了使用鼠标器。鼠标器控制屏幕上的鼠标指针。可以通过在平面上滑动鼠标器，随心所欲地移动鼠标器。当鼠标器将超出范围时，可拿起它再放下。只有当鼠标器接触到一个平面时、鼠标才能移动。在屏幕上移动鼠标指针时文档不受影响，鼠标指针仅起到屏幕定位作用。当你按下鼠标器时，一些事情就在鼠标指针所在的位置上发生了。

Exercises

[Ex. 1] 根据 Text A 回答以下问题。

1) What is cloud broker?

2) What does cloud broker act as?

3) How many essential characteristics do clouds have? What are they?

4) Why are identity management, privacy and access control of particular concern?

5) What is a cloud-oriented architecture related to?

6) What is a ROA based on?

7) How do cloud servers behave? At which level can they be controlled?

8) What were SaaS providers previously referred to? What does the service provider do in the SaaS layer?

9) What is vertical cloud?

10) Who manages the VPC infrastructure? What about the resources allocated to a VPC?

[Ex. 2] 根据 Text B 填空。

1) Cloud storage is a model of ____________ in which the digital data is stored ______________.

2) These cloud storage providers are responsible for keeping the data ______________, and the physical environment ______________.

3) Cloud storage is based on ________________ and is like broader cloud computing in terms of ________________, ________________, ______________, and ______________.

Highly virtualized infrastructure, accessible interfaces, near-instant elasticity and scalability.

4) Organizations can choose between ________________ and ________________ cloud storage options, or a mixture of the two options, depending on relevant decision criteria that is complementary to ______________________________.

5) Cloud storage provides users with immediate access to a broad range of ___________________ hosted in the infrastructure of another organization via __________________.

6) Cloud storage can be mapped as ________________ with the WebDAV protocol. It can function as ________________for organizations with multiple office locations.

7) When data has been distributed it is stored ________________, increasing the risk of ________________to the data.

8) The risk of having data read during transmission can be mitigated through ________________. Encryption in transit ________________ as it is being transmitted to and from the cloud service.

9) Reliability and availability depends on ________________ and on the level of precautions taken by ________________. Reliability should be based on __________ as well as ________________.

10) Users with specific records-keeping requirements, such as public agencies that must __________________ according to statute, may encounter complications with _________________ .

[Ex. 3] 把下列句子翻译为中文。

1) If you choose a different platform, you can still leverage many of the technologies provided here to help you manage your cloud application.

2) Cloud computing allows users to specify the amount of each system resource needed for their application.

3) One of the features of cloud computing is the ability to move applications from one processor environment to another.

4) Many cloud computing security requirements are solvable only by using cryptographic techniques.

5) Management responsibilities are divided between the public cloud provider and the business itself.

6) Governance is the primary responsibility of the owner of a private cloud, and the shared responsibility of the service provider and service consumer in the public cloud.

7) Using a hybrid cloud, organizations can determine the objectives and requirements of the services to be created, and obtain them based on the most suitable alternative.

8) One limitation of this monitoring suite is that it is available only as hosted service and not as

standalone application.

9) With the SaaS model, you can reduce IT costs because you no longer need to support multiple platforms and versions.

10) A virtual network gives a customer a common LAN—like network where the servers can be located in the physical data center (virtualized or not), private cloud, and public cloud.

[Ex. 4] 软件水平考试真题自测。

从供选择的答案中选出应填入下列英语文句中____内的正确答案，把编号写在答卷的对应栏内。

Packet-switching wireless networks are preferable __A__ when transmissions are __B__ because of the way charges are __C__ per packet. Circuit-switched networks are preferable for transferring large files or for other lengthy transmissions because customers are __D__ for the __E__ of time they use the network.

供选择的答案：

A：①to ②for ③than ④only

B：①long ②short ③large ④small

C：①computing ②incurious ③incurved ④incurred

D：①charged ②fined ③free ④controlled

E：①point ②start ③length ④end

[Ex. 5] 听短文填空。

Exercises 9

Cloud platform as a service refers to the delivery of a computing platform solution as a service. Synonymous with PaaS (platform as a service), cloud platform as a service delivery ____1____ a complete framework that facilitates the development, deployment and ____2____ of on-demand software, which is also known as SaaS (software as a service) or applications as a service. Some PaaS solutions provide an open ____3____ that allows the use of any programming language, database, operating system or server.

There are essentially two types of PaaS ____4____ : application delivery only PaaS solutions and complete PaaS solutions. The former only delivers the bare bones functionality that is needed to ____5____ applications in the cloud. Facilities for developing, debugging or testing SaaS solutions are not available with this type of PaaS solution. The later provides a robust computing platform with a full solution ____6____ that enables the efficient management and delivery of on demand software.

Complete PaaS solutions provide a platform for SaaS development, SaaS deployment and SaaS testing, in support of the full application life ____7____ . A complete PaaS solution may also ____8____ the following capabilities: team collaboration tools, security management tools, database integration tools, application version management and more. On a detailed level, a complete PaaS solution includes the following ____9____ : user on-boarding or provisioning, subscription management, billing & metering, payment processing, reporting & analytics and more.

Apprenda's private PaaS offering is an example of a complete PaaS solution that delivers a full set of features that fully ___10___ the operational requirements of SaaS solution providers or independent software vendors.

Reference Translation

云计算词汇

CDN（内容分发网络）

由包含数据副本的多台计算机组成的系统，这些数据位于网络不同的地方，这样客户端可以访问最近的副本。

云

喻指一个全球性的网络，起初指电话网，现在常用来代表因特网。

云应用

没有安装在本地机器上的软件应用程序——它总是通过因特网访问。云金字塔“顶”层的应用程序通过 Web 浏览器运行并与之交互。云应用受到严格控制并很少进行修改，例如，Gmail 或 SalesForce.com 等。

云即服务（CaaS）

云即服务是一个云计算服务，它已经在一个平台上开放且其他人可以在此平台进行搭建。

云桥

云桥以这样一种方式运行应用程序：其组件被集成到多个云环境中（可能是内部/私有云和外部/公共云的任意组合）。

云经纪人

云经纪人是创建和维护多个云服务供应商的关系的实体。它可以作为云服务的客户和提供商之间的联络人，为每个客户选择最佳的供应商和监管服务。

云破裂

云破裂是指云发生中断或安全漏洞以及出现数据不可用的情况。

云中心

云中心是指在云中架构的一个数据中心，利用基于标准的虚拟化组件作为类似数据中心的基础设施；例如，一个大公司（如亚马孙）租用云中心的基础设施。

云计算

云计算是一种计算能力，它提供了抽象的计算资源及其基础技术架构（例如，服务器、存储器、网络），对可快速供应和释放的许多可配置的共享计算资源能实现便捷的、按需的网络访问，只需最少的管理或服务提供商交互。这个定义表示云有 5 个基本特征：按需自助服务、广泛的网络接入、资源池、快速弹性和定制服务。狭义的云计算是抽象了服务器详细信息的客户——服务器计算；客户请求的是一个服务（资源），而不是特定的服务器（计算机）。云计算实现了基础设施即服务（IaaS）、平台即服务（PaaS）和软件即服务（SaaS）。云计算意味着你可以得到基础设施、应用程序和业务流程作为一个因特网（或者你自己的网络）服务。

云计算安全

云计算安全是一套基于控制的技术和策略，用来保护云计算所用的信息、数据应用程序和基础设施。

因为云计算的本质是共享资源，因此对身份管理、隐私和访问控制尤为关注。随着使用云计算的组织和进行数据操作的相关云提供商的增多，在这些方面和其他易受攻击领域的安全已经成为组织与云计算提供商签约时优先考虑的事项。

云计算安全过程应该解决云提供商的安全控制问题，以保护客户的数据安全、隐私和遵守必要的法规。在出现云安全漏洞时，该过程也能保证业务的连续性和数据的备份计划。

云托管

云托管是一种互联网托管，客户按需租用虚拟化、动态可扩展的基础设施。使用者常常可以选择操作系统和其他基础设施组件。典型的云托管是自助服务，按照小时或按月计费，并通过 Web 接口或 API 来控制。

云基础设施

云金字塔的“底”层（或基础层）是通过半虚拟化交付计算机基础设施。这些基础设施包括服务器、网络和其他以基础设施 Web 服务或“云中心”形式所交付的硬件设备。在这个级别提供对基础设施的全面控制，例如，GoGrid 或 Amazon Web 服务。

云操作系统

云操作系统是专门为在提供商数据中心运行而设计的计算机操作系统，并通过因特网或其他网络交付给用户。Windows Azure 是云操作系统的一个示例。云操作系统有时也指基于云的客户端操作系统，如谷歌的 Chrome 操作系统。

面向云的体系结构（COA）

面向云的体系结构（COA）是一个概念性模型，包括了云环境中的所有元素。在信息技术领域，体系结构是指信息系统的整体结构和组成该系统的实体之间的关系。

面向云的体系结构与面向服务的体系结构（SOA）和事件驱动体系结构（EDA）均相关，也组合了面向资源的体系结构（ROA）和面向超媒体的体系结构（HOA）两个结构模型。ROA 基于这样的想法：任何一个可以分配到统一资源标识符（URI）的实体都是资源。因此，资源不仅包括基础设施元素（如服务器、计算机和其他设备），也包括网页、脚本和 JSP/ASP 页面以及其他实体。超媒体扩展了超文本链接的概念，包括了任何一组多媒体对象的链接（包括声音、视频和虚拟现实）。

云平台

云平台在云金字塔的“中间”层，它提供一个计算平台或框架（例如，.NET、Ruby on Rails 或 Python）作为服务或堆栈。只对该平台或框架进行控制，但不是在一个较低的层级（服务器基础设施）上，例如，谷歌的 AppEngine 或微软的 Azure。

云可移植性

云可移植性是指在来自不同的云提供商的云计算环境中迁移应用程序（通常也包括其关联的数据），也可以在私有云与公有云、内部云与外部云之间迁移应用程序。

云提供商

该公司为其他组织和/或个人提供基于云计算的平台、基础设施、应用程序或存储服务，这些通常都要收费。

云金字塔

云金字塔按照功能分层，是对云计算层级的直观表示。其简化的版本包括三层：基础设施层、平台层和应用层。

云服务器

云服务器是运行 Windows 或 Linux 操作系统的虚拟化服务器，通过 Web 接口或 API 而具体化。云服务器的行为与物理服务器一样，可以由管理员或根级别用户进行控制，具体由服务器类型和云托管服务提供商来决定。

云标准

标准是商定一致的做某一事情的办法。云标准确保了互操作性，这样你就能把工具、应用程序、虚拟图像等用到其他云环境中，而无须重做任何事情。云标准的便携性使你可以把一个应用程序或一个供应商的实施实例在另一个供应商处顺利运行。

云存储

云存储是一种服务，允许客户通过因特网或其他网络传输数据，并将数据保存在第三方维护的异地存储系统中。

云件

云件是一个通用术语，指各种各样的软件（通常在基础设施层）。它能够构建、部署、运行或管理云计算环境中的应用程序。

基于消费的定价模型

基于消费的定价模型是一种定价模式，即根据客户消费的服务量收费，而不是根据时间。例如，一个云存储提供商可以根据存储信息所需的吉字节数收费。

客户自助服务

客户自助服务是指客户能够自己提供、管理和终止服务，而无须服务供应商参与，通过一个 Web 接口或程序来调用服务的 API。

弹性和可扩展性

云是弹性的，即可以按照需求分配资源的大小。弹性产生了可扩展性，即云可以根据需求高峰和低谷来调高和调低资源规模。可扩展性也意味着在用户增加和需求变化时可调整应用程序。

外部云

外部云是指由组织之外的第三方提供的公有或私有云服务。云计算环境是在组织边界之外的。

托管应用程序

托管应用程序是指运行在远程服务器的基于因特网或基于 Web 的应用软件程序，它可以通过连接因特网的个人计算机或瘦客户端进行访问。

混合云

混合云是一种整合了多个内部和/或外部供应商的网络环境，组合了公有云和私有云的多个方面。

身份管理

管理个人身份信息，以便合理管理对计算机资源、应用程序、数据和服务的访问。

基础设施即服务（IaaS）

云基础设施服务或“基础设施即服务”（IaaS）把计算机基础设施作为服务来提供，通常是一个平台虚拟化环境。客户不是购买服务器、软件、数据中心空间或网络设备，而是将这些资源作为完全外包的服务来购买。该服务通常以效用计算为基础来计费，所消耗的资源量（和因此而产生的费用）通常会反映活动的等级。它是网络托管和虚拟专用服务产品的发展。

内部云

内部云是一种私有云，其服务由自己组织中的 IT 部门提供。

按需服务

按需服务是一种客户可根据需要购买云服务的模型，例如，在项目持续期，客户可以根据需要购买额外的服务器，待该项目完成后再回到以前的水平。

个人云

个人云是 MiFi（移动无线保真）的同义词，是一种个人无线路由器。它接收一个移动无线数据信号并将其转换为 WiFi。

平台即服务（PaaS）

平台即服务，也就是云平台服务，是指计算平台（操作系统和相关服务）作为服务由供应商通过互联网提供。PaaS 层提供黑箱服务，使得开发人员可以在计算基础设施之上构建应用程序。它可能包括用于构建服务的各种开发工具或数据访问和数据库服务，或收费服务。

私有云

私有云是指安装在公司防火墙里的虚拟云数据中心。它也可以是一个云提供商数据中心中专用于公司的私有空间，还可以是组织的防火墙之后的内部云。该公司的 IT 部门为客户提供软件和硬件服务——客户就是那些为公司工作的人。供应商喜欢“私有云”这个词。

公有云

公有云是通过公共因特网提供的服务，任何想要购买该服务的人都可使用。

软件即服务（SaaS）

云应用服务，即供应商通过因特网提供应用程序，因此客户不必购买应用程序，也不必安装和运行在自己的计算机上。SaaS 提供商以前被称为 ASP（应用服务提供商）。在 SaaS 层，服务提供商托管软件，因此客户不需要安装、管理该软件或者为它购买硬件，只需要连接并使用即可。客户关系管理作为一种服务就是 SaaS。

提供自助服务

云客户可以自助提供云服务，而无须经历一个漫长的过程。客户请求的计算、存储、软件、进程或更多东西都来自服务提供商。在使用这些资源后，可以自动取消这些资源的供应。

标准化的接口

云服务应具有标准化的 API，这些 API 提供两个应用程序或数据源间相互通信的指令。一个标准化的接口让客户更方便地联结云服务。

垂直云

垂直云是为用于特定行业而优化的云计算环境，如用于医疗保健或金融服务。

虚拟私有云（VPC）

术语“虚拟私有云”（VPC）类似于也来自于人们熟悉的虚拟专用网络（VPN），但被应用于云计算之中。它是把公有云转换成一个虚拟私有云的概念。

VPC 是云计算的一个混合模式。这种模式在一个公有云提供商的基础设施中提供私有云解决方案。

VPC 是一种云计算服务，公有云提供商分离出公有云基础设施的一部分，供私人使用。VPC 的基础设施由公有云提供商管理；然而，任何其他客户不能共享分配给 VPC 的资源。

Unit 10

Text A

Big Data

1. Definition

Big data is a general term used to describe the voluminous amount of unstructured and semi-structured data a company creates—data that would take too much time and cost too much money to load into a relational database for analysis. Although big data doesn't refer to any specific quantity, the term is often used when speaking about petabytes and exabytes of data.

Text A
Big Data

A primary goal for looking at big data is to discover repeatable business patterns. It's generally accepted that unstructured data, most of which is located in text files, accounts for at least 80% of an organization's data. If left unmanaged, the sheer volume of unstructured data that's generated each year within an enterprise can be costly in terms of storage. Unmanaged data can also pose a liability if information cannot be located in the event of a compliance audit or lawsuit.

Big data analytics is often associated with cloud computing because the analysis of large data sets in real-time requires a framework like MapReduce to distribute the work among tens, hundreds or even thousands of computers.

2. Big data analytics

Big data analytics is the process of examining large amounts of data of a variety of types to uncover hidden patterns, unknown correlations and other useful information. Such information can provide competitive advantages over rival organizations and result in business benefits, such as more effective marketing and increased revenue.

The primary goal of big data analytics is to help companies make better business decisions by enabling data scientists and other users to analyze huge volumes of transaction data as well as other data sources that may be left untapped by conventional business intelligence (BI) programs.[1] These other data sources may include Web server logs and Internet clickstream data, social media activity reports, mobile-phone call detail records and information captured by sensors. Some people exclusively associate big data and big data analytics with unstructured data of that sort, but consulting firms like Gartner Inc. and Forrester Research Inc. also consider transactions and other structured data to be valid forms of big data.

Big data analytics can be done with the software tools commonly used as part of advanced analytics disciplines such as predictive analytics and data mining. But the unstructured data sources used for big data analytics may not fit in traditional data warehouses. Furthermore, traditional data warehouses may not be able to handle the processing demands posed by big data. As a result, a new class of big data technology has emerged and is being used in many big data analytics environments. The technologies associated with big data analytics include NoSQL databases, Hadoop and MapReduce. These technologies form the core of an open source software framework that supports the processing of large data sets across clustered systems.

Potential pitfalls that can trip up organizations on big data analytics initiatives include a lack of internal analytics skills and the high cost of hiring experienced analytics professionals, plus challenges in integrating Hadoop systems and data warehouses, although vendors are starting to offer software connectors between those technologies.[2]

3. Big data management

Big data management is the organization, administration and governance of large volumes of both structured and unstructured data.

The goal of big data management is to ensure a high level of data quality and accessibility for business intelligence and big data analytics applications. Corporations, government agencies and other organizations employ big data management strategies to help them contend with fast-growing pools of data, typically involving many terabytes or even petabytes of information saved in a variety of file formats. Effective big data management helps companies locate valuable information in large sets of unstructured data and semi-structured data from a variety of sources, including call detail records, system logs and social media sites.

Most big data environments go beyond relational databases and traditional data warehouse platforms to incorporate technologies that are suited to processing and storing nontransactional forms of data. The increasing focus on collecting and analyzing big data is shaping new platforms that combine the traditional data warehouse with big data systems in a logical data warehousing architecture. As part of the process, they must decide what data must be kept for compliance reasons, what data can be disposed of and what data should be kept and analyzed in order to improve current business processes or provide a business with a competitive advantage. This process requires careful data classification so that ultimately, smaller sets of data can be analyzed quickly and productively.

4. Big data as a service (BDaaS)

Big data as a service (BDaaS) is the delivery of statistical analysis tools or information by an outside provider that helps organizations understand and use insights gained from large information sets in order to gain a competitive advantage.[3]

Given the immense amount of unstructured data generated on a regular basis, BDaaS is intended to free up organizational resources by taking advantage of the predictive analytics skills of an outside provider to manage and assess large data sets, rather than hiring in-house staff for those functions.[4] It can take the form of software that assists with data processing or a contract for the services of a team of data scientists.

BDaaS is a form of managed services, similar to software as a service or infrastructure as a service. It often relies upon cloud storage to preserve continual data access for the organization that owns the information as well as the provider working with it.

5. Unstructured data

Unstructured data is a generic label for describing any corporate information that is not in a database. Unstructured data can be textual or non-textual. Textual unstructured data is generated in media like E-mail messages, PowerPoint presentations, Word documents, collaboration software and instant messages. Non-textual unstructured data is generated in media like JPEG images, MP3 audio files and Flash video files.

The information contained in unstructured data is not always easy to locate. It requires that data in both electronic and hard copy documents and other media be scanned, so a search application can parse out concepts based on words used in specific contexts.[5] This is called semantic search. It is also referred to as enterprise search.

In customer-facing businesses, the information contained in unstructured data can be analyzed to improve customer relationship management and relationship marketing. As social media applications like Twitter and Facebook go mainstream, the growth of unstructured data is expected to far outpace the growth of structured data.

6. Data Mining

Generally, data mining (sometimes called data or knowledge discovery) is the process of analyzing data from different perspectives and summarizing it into useful information—information that can be used to increase revenue, cuts costs, or both. Data mining software is one of a number of analytical tools for analyzing data. It allows users to analyze data from many different dimensions or angles, categorize it, and summarize the relationships identified. Technically, data mining is the process of finding correlations or patterns among dozens of fields in large relational databases.

Data mining parameters include:

Association—looking for patterns where one event is connected to another event.

Sequence or path analysis—looking for patterns where one event leads to another later event.

Classification—looking for new patterns (May result in a change in the way the data is organized but that's OK).

Clustering—finding and visually documenting groups of facts not previously known.

Forecasting—discovering patterns in data that can lead to reasonable predictions about the future (This area of data mining is known as predictive analytics).

Data mining techniques are used in many research areas, including mathematics, cybernetics, genetics and marketing. Web mining, a type of data mining used in customer relationship management (CRM), takes advantage of the huge amount of information gathered by a Web site to look for patterns in user behavior.

7. Data warehouse

A data warehouse is a central repository for all or significant parts of the data that an enterprise's various business systems collect. The term was coined by W. H. Inmon. IBM sometimes uses the term"information warehouse".

Typically, a data warehouse is housed on an enterprise mainframe server. Data from various online transaction processing (OLTP) applications and other sources is selectively extracted and organized on the data warehouse database for use by analytical applications and user queries. Data warehousing emphasizes the capture of data from diverse sources for useful analysis and access, but does not generally start from the point-of-view of the end user or knowledge worker who may need access to specialized, sometimes local databases. The latter idea is known as the data mart.

Applications of data warehouses include data mining, Web mining and decision support systems (DSS).

New Words

voluminous	[və'lu:mɪnəs]	*adj.*体积大的，庞大的
semi-structured	['semɪ-'strʌktʃəd]	*adj.*半结构化的
quantity	['kwɒntətɪ]	*n.*量，数量
petabyte	['petəbait]	*n.*千万亿字节，2^{50}字节
exabyte	['eksəbaɪt]	*n.*艾字节，2^{60}字节
discover	[dɪ'skʌvə]	*vt.*发现，发觉
repeatable	[rɪ'pi:təbl]	*adj.*可重复的
pattern	['pætən]	*n.*式样，模式
unmanage	[ʌn'mænɪdʒ]	*vi.*不处理，应付过去 *vt.*不管理，不控制，不操纵，不维持
sheer	[ʃɪə]	*adj.*全然的，纯粹的，绝对的，彻底的
liability	[ˌlaɪə'bɪlətɪ]	*n.*责任，义务，倾向，债务
analytics	[ˌænə'lɪtɪks]	*n.*分析学，解析学，分析论
uncover	[ʌn'kʌvə]	*vt.*揭开，揭露，揭示
correlation	[ˌkɒrə'leɪʃn]	*n.*相互关系，相关(性)
competitive	[kəm'petətɪv]	*adj.*竞争的
rival	['raɪvl]	*n.*竞争者，对手 *v.*竞争，对抗，相匹敌
revenue	['revənju:]	*n.*收入，税收
huge	[hju:dʒ]	*adj.*巨大的，极大的
untapped	[ˌʌn'tæpt]	*adj.*未使用的
log	[lɒg]	*n.*日志
clickstream	['klɪkstri:m]	*n.*点击流量
sensor	['sensə]	*n.*传感器
exclusively	[ɪk'sklu:sɪvlɪ]	*adv.*专有地，排外地
valid	['vælɪd]	*adj.*有效的，有根据的，正当的
predictive	[prɪ'dɪktɪv]	*adj.*预言性的
warehouse	['weəhaʊs]	*n.*仓库 *vt.*储入仓库
cluster	['klʌstə]	*n.*串，丛

		*vi.*丛生，成群
potential	[pə'tenʃl]	*adj.*潜在的，可能的
pitfall	['pɪtfɔ:l]	*n.*潜在的困难；陷阱，圈套，诱惑
accessibility	[əkˌsesə'bɪlətɪ]	*n.*可访问性
strategy	['strætədʒɪ]	*n.*策略
format	['fɔ:mæt]	*n.*格式
		*vt.*格式化(磁盘)
valuable	['væljuəbl]	*adj.*有价值的
site	[saɪt]	*n.*网站
collect	[kə'lekt]	*v.*收集，聚集，集中，搜集
dispose	[dɪ'spəʊz]	*v.*处理
careful	['keəfl]	*adj.*小心的，仔细的
immense	[ɪ'mens]	*adj.*极广大的，无边的
regular	['regjələ]	*adj.*规则的，有秩序的，合格的，定期的
preserve	[prɪ'zɜ:v]	*vt.*保存，保持，保护
label	['leɪbl]	*vt.*贴标签于，指……为，分类，标注
		*n.*标签，标志
textual	['tekstʃuəl]	*adj.*本文的
scan	[skæn]	*v.&n.*扫描
mainstream	['meɪnstri:m]	*n.*主流
outpace	[ˌaʊt'peɪs]	*vt.*超过……速度，赶过
perspective	[pə'spektɪv]	*n.*观点，看法，观点，观察
dimension	[daɪ'menʃn]	*n.*尺度，维(数)
association	[əˌsəʊʃɪ'eɪʃn]	*n.*协作，联合
clustering	['klʌstərɪŋ]	*n.*聚类
forecast	['fɔ:kɑ:st]	*n.&vt.*预测，预报
significant	[sɪg'nɪfɪkənt]	*adj.*有意义的，重大的，重要的
selective	[sɪ'lektɪv]	*adj.*选择的，选择性的
extract	[ɪk'strækt]	*vt.*提取，析取

Phrases

big data	大数据
relational database	关系数据库
text file	文本文件
data set	数据集
transaction data	业务数据；事务数据；交易数据
social media	社交媒体
consulting firm	咨询公司
data warehouse	数据仓库

trip up	犯错误，绊倒，使摔倒
data classification	数据分类
competitive advantage	竞争优势
Software as a Service	软件即服务
Infrastructure as a Service	基础架构即服务
cloud storage	云存储
instant message	即时消息，即时报文
semantic search	语义搜索
enterprise search	企业搜索
relationship marketing	关系营销
data mining	数据挖掘
knowledge discovery	知识发现
data mart	数据集市
Web Mining	网络挖掘

Abbreviations

BI (Business Intelligence)	商务智能，商业智能
BDaaS (Big data as a service)	大数据即服务
JPEG (Joint Photographic Experts Group)	联合图像专家组
OLTP (OnLine Transaction Processing)	联机事务处理
DSS (Decision Support System)	决策支持系统

Notes

[1] The primary goal of big data analytics is to help companies make better business decisions by enabling data scientists and other users to analyze huge volumes of transaction data as well as other data sources that may be left untapped by conventional business intelligence (BI) programs.

本句中，to help companies make better business decisions 作表语，和它前面的 is 一起作谓语。by 引导一个方式状语，修饰谓语。that may be left untapped by conventional business intelligence (BI) programs 是一个定语从句，修饰和限定 other data sources。

[2] Potential pitfalls that can trip up organizations on big data analytics initiatives include a lack of internal analytics skills and the high cost of hiring experienced analytics professionals, plus challenges in integrating Hadoop systems and data warehouses, although vendors are starting to offer software connectors between those technologies.

本句中，that can trip up organizations on big data analytics initiatives 是一个定语从句，修饰和限定 Potential pitfalls。although vendors are starting to offer software connectors between those technologies 是一个让步状语从句。

[3] Big data as a service (BDaaS) is the delivery of statistical analysis tools or information by an outside provider that helps organizations understand and use insights gained from large information sets in order to gain a competitive advantage.

本句中，that helps organizations understand and use insights gained from large information sets in order to gain a competitive advantage 是一个定语从句，修饰和限定 an outside provider。在该从句中，gained from large information sets 是一个过去分词短语做定语，修饰和限定 insights。in order to gain a competitive advantage 作目的状语。

[4] Given the immense amount of unstructured data generated on a regular basis, BDaaS is intended to free up organizational resources by taking advantage of the predictive analytics skills of an outside provider to manage and assess large data sets, rather than hiring in-house staff for those functions.

本句中，Given the immense amount of unstructured data generated on a regular basis 作条件状语；generated on a regular basis 是一个过去分词短语作定语，修饰和限定 unstructured data。by taking advantage of the predictive analytics skills of an outside provider to manage and assess large data sets, rather than hiring in-house staff for those functions 作方式状语；to manage and assess large data sets 是一个动词不定式短语，作目的状语；rather than 的意思是“而不是”。

[5] It requires that data in both electronic and hard copy documents and other media be scanned, so a search application can parse out concepts based on words used in specific contexts.

本句中，that data in both electronic and hard copy documents and other media be scanned 是一个宾语从句；在该从句中，be scanned 前面省略了 should。so a search application can parse out concepts based on words used in specific contexts 是一个目的状语；在该从句中，based on words 是一个过去分词短语，作定语，修饰和限定 concepts；used in specific contexts 也是一个过去分词短语，作定语，修饰和限定 words。

Text B

Data Mining

Data mining is a powerful new technology with great potential to help companies focus on the most important information in the data they have collected about the behavior of their customers and potential customers. It discovers information within the data that queries and reports can't effectively reveal.

1. What is data mining?

Data mining, or knowledge discovery, is the computer-assisted process of digging through and analyzing enormous sets of data and then extracting the meaning of the data. Data mining tools predict behaviors and future trends, allowing businesses to make proactive, knowledge-driven decisions. Data mining tools can answer business questions that traditionally were too time-consuming to resolve. They scour databases for hidden patterns, finding predictive information that experts may miss because it lies outside their expectations.

Data mining derives its name from the similarities between searching for valuable information in a large database and mining a mountain for a vein of valuable ore. Both processes require either sifting through an immense amount of material, or intelligently probing it to find where the value resides.

2. What can data mining do?

Although data mining is still in its infancy, companies in a wide range of industries—including retail, finance, heath care, manufacturing transportation, and aerospace—are already using data mining tools and techniques to take advantage of historical data. By using pattern recognition technologies and statistical and mathematical techniques to sift through warehoused information, data mining helps analysts recognize significant facts, relationships, trends, patterns, exceptions and anomalies that might otherwise go unnoticed.

For businesses, data mining is used to discover patterns and relationships in the data in order to help make better business decisions. Data mining can help spot sales trends, develop smarter marketing campaigns, and accurately predict customer loyalty. Specific uses of data mining include:

- Market segmentation—identify the common characteristics of customers who buy the same products from your company.
- Customer churn—predict which customers are likely to leave your company and go to a competitor.
- Fraud detection—identify which transactions are most likely to be fraudulent.
- Direct marketing—identify which prospects should be included in a mailing list to obtain the highest response rate.
- Interactive marketing—predict what each individual accessing a Web site is most likely interested in seeing.
- Market basket analysis—understand what products or services are commonly purchased together; e.g., beer and diapers.
- Trend analysis—reveal the difference between a typical customer this and last month.

Data mining technology can generate new business opportunities by:

Automated prediction of trends and behaviors: data mining automates the process of finding predictive information in a large database. Questions that traditionally required extensive hands-on analysis can now be directly answered from the data. A typical example of a predictive problem is targeted marketing. Data mining uses data on past promotional mailings to identify the targets most likely to maximize return on investment in future mailings. Other predictive problems include forecasting bankruptcy and other forms of default, and identifying segments of a population likely to respond similarly to given events.

Automated discovery of previously unknown patterns: data mining tools sweep through databases and identify previously hidden patterns. An example of pattern discovery is the analysis of retail sales data to identify seemingly unrelated products that are often purchased together. Other pattern discovery problems include detecting fraudulent credit card transactions and identifying anomalous data that could represent data entry keying errors.

Using massively parallel computers, companies dig through volumes of data to discover patterns about their customers and products. For example, grocery chains have found that when men go to a supermarket to buy diapers, they sometimes walk out with a six-pack of beer as well. Using that information, it's possible to lay out a store so that these items are closer.

AT&T, A.C. Nielson, and American Express are among the growing ranks of companies implementing data mining techniques for sales and marketing. These systems are crunching through terabytes of point-of-sale data to aid analysts in understanding consumer behavior and promotional strategies. Why? To gain a competitive advantage and increase profitability!

Similarly, financial analysts are plowing through vast sets of financial records, data feeds, and other information sources in order to make investment decisions. Health-care organizations are examining medical records to understand trends of the past so they can reduce costs in the future.

3. How data mining works

How is data mining able to tell you important things that you didn't know or what is going to happen next? The technique that is used to perform these feats is called modeling. Modeling is simply the act of building a model (a set of examples or a mathematical relationship) based on data from situations where the answer is known and then applying the model to other situations where the answers aren't known. Modeling techniques have been around for centuries, of course, but it is only recently that data storage and communication capabilities required to collect and store huge amounts of data, and the computational power to automate modeling techniques to work directly on the data have been available.

As a simple example of building a model, consider the director of marketing for a telecommunications company. He would like to focus his marketing and sales efforts on segments of the population most likely to become big users of long distance services. He knows a lot about his customers, but it is impossible to discern the common characteristics of his best customers because there are so many variables. From his existing database of customers, which contains information such as age, sex, credit history, income, zip code, occupation, etc., he can use data mining tools, such as neural networks, to identify the characteristics of those customers who make lots of long distance calls. For instance, he might learn that his best customers are unmarried females between the age of 34 and 42 who make in excess of $60,000 per year. This, then, is his model for high value customers, and he would budget his marketing efforts accordingly.

4. Data mining technologies

The analytical techniques used in data mining are often well-known mathematical algorithms and techniques. What is new is the application of those techniques to general business problems made possible by the increased availability of data and inexpensive storage and processing power. Also, the use of graphical interfaces has led to tools becoming available that business experts can easily use.

Some of the tools used for data mining are:

Artificial neural networks—non-linear predictive models that learn through training and resemble biological neural networks in structure.

Decision trees—tree-shaped structures that represent sets of decisions. These decisions generate rules for the classification of a dataset.

Rule induction—the extraction of useful if-then rules from data based on statistical significance.

Genetic algorithms—optimization techniques based on the concepts of genetic combination, mutation, and natural selection.

Nearest neighbor—a classification technique that classifies each record based on the records most similar to it in an historical database.

5. Real-world examples

Details about who calls whom, how long they are on the phone, and whether a line is used for fax as well as voice can be invaluable in targeting sales of services and equipment to specific customers. But these tidbits are buried in masses of numbers in the database. By delving into its extensive customer-call database to manage its communications network, a regional telephone company identifies new types of unmet customer needs. Using its data mining system, it discovers how to pinpoint prospects for additional services by measuring daily household usage for selected periods. For example, households that make many lengthy calls between 3 p.m. and 6 p.m. are likely to include teenagers who are prime candidates for their own phones and lines. When the company uses target marketing that emphasizes convenience and value for adults–"Is the phone always tied up?" –hidden demand surfaces. Extensive telephone use between 9 a.m. and 5 p.m. characterized by patterns related to voice, fax, and modem usage suggests a customer has business activity. Target marketing offering those customers"business communications capabilities for small budgets"results in sales of additional lines, functions, and equipment.

The ability to accurately gauge customer response to changes in business rules is a powerful competitive advantage. A bank searching for new ways to increase revenues from its credit card operations tested a nonintuitive possibility: would credit card usage and interest earned increase significantly if the bank halved its minimum required payment? With hundreds of gigabytes of data representing two years of average credit card balances, payment amounts, payment timeliness, credit limit usage, and other key parameters, the bank used a powerful data mining system to model the impact of the proposed policy change on specific customer categories. The bank discovered that cutting minimum payment requirements for small, targeted customer categories could increase average balances and extend indebtedness periods, generating more than $25 million in additional interest earned. Merck-Medco Managed Care is a mail-order business which sells drugs to the country's largest health care providers. Merck-Medco is mining its one terabyte data warehouse to uncover hidden links between illnesses and known drug treatments, and spot trends that help pinpoint which drugs are the most effective for what types of patients. The results are more effective treatments that are also less costly. Merck-Medco's data mining project has helped customers save an average of 10%～15% on prescription costs.

6. The future of data mining

In the short-term, the results of data mining will be in profitable business related areas. Micro-marketing campaigns will explore new niches. Advertising will target potential customers with new precision.

In the medium term, data mining may be as common and easy to use as E-mail. We may use these tools to find the best airfare to New York, root out a phone number of a long-lost classmate, or find the best prices on lawn mowers.

The long-term prospects are truly exciting. Imagine intelligent agents turning loose on medical research data or on sub-atomic particle data. Computers may reveal new treatments for diseases.

New Words

behavior	[bɪ'heɪvjə]	*n.*举止，行为
dig	[dɪg]	*v.*掘，挖，搜集
proactive	[ˌprəʊ'æktɪv]	*adj.*积极的，主动地
time-consuming	[taɪm-kən'sju:mɪŋ]	*adj.*耗费时间的，旷日持久的
scour	['skaʊə]	*v.*四处搜集，冲洗，擦亮
vein	[veɪn]	*n.*矿脉，纹理
probe	[prəʊb]	*v.*探查，探测
transportation	[ˌtrænspɔ:'teɪʃn]	*n.*运输，运送
aerospace	['eərəʊspeɪs]	*n.*航空航天
sift	[sɪft]	*vt.*筛分，精选；审查 *vi.*筛，细查
unnoticed	[ˌʌn'nəʊtɪst]	*adj.*不引人注意的，被忽视的
spot	[spɒt]	*vt.*认出，发现
churn	[tʃɜ:n]	*v.*流失
fraudulent	['frɔ:djələnt]	*adj.*欺诈的，欺骗性的
bankruptcy	['bæŋkrʌptsi]	*n.*破产
sweep	[swi:p]	*v.*扫过，掠过
anomalous	[ə'nɒmələs]	*adj.*不规则的，反常的
grocery	['grəʊsəri]	*n.*食品杂货店，食品店，杂货铺
crunch	[krʌntʃ]	*v.*嘎扎嘎扎地咬嚼，压碎，扎扎地踏过
feat	[fi:t]	*n.*技艺；功绩，壮举
discern	[dɪ'sɜ:n]	*v.*认识，洞悉，辨别，看清楚
budget	['bʌdʒɪt]	*n.*预算 *vi.*做预算，编入预算
artificial	[ˌɑ:tɪ'fɪʃl]	*adj.*人造的
non-linear	[nɒn-'lɪnɪə]	*adj.*非线性的
induction	[ɪn'dʌkʃn]	*n.*归纳法
optimization	[ˌɒptɪmaɪ'zeɪʃən]	*n.*最佳化，最优化
mutation	[mju:'teɪʃn]	*n.*变化，转变，（生物物种的）突变
invaluable	[ɪn'væljuəbl]	*adj.*无价的，价值无法衡量的
tidbit	['tɪdbɪt]	*n.*珍品，珍闻
bury	['beri]	*vt.*掩埋，隐藏
unmet	[ˌʌn'met]	*adj.*未满足的，未相遇的，未应付的
pinpoint	['pɪnpɔɪnt]	*n.*精确 *adj.*极微小的 *v.*查明

non-intuitive	[nɒnɪn'tju: ɪtɪv]	*adj.*非直觉的
possibility	[ˌpɒsə'bɪlətɪ]	*n.*可能性
earn	[ɜ:n]	*v.*赚得，获得
halve	[hɑ:v]	*vt.*二等分，平分，分享，减半
indebtedness	[ɪn'detɪdnəs]	*n.*亏欠，债务
mail-order	[ˌmeɪl-'ɔ:də]	*adj.*邮购的
drug	[drʌg]	*n.*药，麻药，毒品 *vi.*吸毒 *vt.*使服毒品，毒化
treatment	['tri:tmənt]	*n.*处理，治疗
prescription	[prɪ'skrɪpʃn]	*n.*处方，药方
profitable	['prɒfɪtəbl]	*adj.*有利可图的
niche	[nɪtʃ]	*n.*小生态环境，商机

Phrases

computer-aided process	计算机辅助过程
knowledge-driven decision	知识驱动决策
sift through	筛选
heath care	卫生保健
pattern recognition	模式识别
business decision	业务决策，商务决定
customer churn	客户流失
response rate	响应率
market basket analysis	购物篮分析
business opportunity	业务机会，商业机会
targeted marketing	目标市场
seemingly unrelated product	似乎无关的产品
point-of-sale data	销售终端数据
investment decision	投资决策
neural network	神经网络
graphical interface	图形界面，图形接口
predictive model	预测模型
decision tree	决策树，判定树
rule induction	规则归纳
genetic algorithm	遗传算法
nearest neighbor(algorithm)	最邻近算法
delve into	钻研，深入研究
root out	搜寻
lawn mower	割草机，剪草机

Word Building

常用前缀列举如下：

1. -ivity 表示：性质；情况；状态

act 做，活动——activity 活动性，活动

productive 有生产能力的，多产的——productivity 生产力，生产率

sensitive 敏感的，灵敏的——sensitivity 敏感性，灵敏度

passive 被动的——passivity 被动性

2. -ment 表示：行为；状态；过程

develop 发展，开发——development 发展，开发

agree 同意——agreement 同意，协议

equipment 装备，配备——equipment 设备

invest 投资——investment 投资

require 要求，需要——requirement 需要

adjust 调整——adjustment 调整

3. -ness 表示：性质；情况；状语

ill 病，疾病——illness 病，疾病

firm 坚定的，牢固的——firmness 结实，坚定

4. -or 表示：……的人

edit 编辑——editor 编辑

invent 发明——inventor 发明者

visit 参观，访问——visitor 来访者

5. -ship 表示：情况；技能；身份；职位；极限

friend 朋友，友人——friendship 友谊，友好

member 成员——membership 成员资格

professor 教授——professorship 教授身份

owner 物主，拥有者——ownership 所有权，所有制

6. -th 表示：动作；过程；状态；性质

grow 成长，生长，增长——growth 成长，生长，增长

strong 强壮的——strength 力量

deep 深的——depth 深度

7. -tion 表示：行为的过程，结果，状况

opt 选择——option 选择，选择权

add 增加——addition 增加

eliminate 消灭，排除——elimination 消灭，排除

execute 完成，执行——execution 完成，执行

suggest 建议——suggestion 建议

situate 位于——situation 位置，处境

8. -ty 表示：性质，状态，情况，构成抽象名词
safe 安全的——safety 安全
beautiful 美丽的，漂亮的——beauty 美丽
cruel 残忍的，残酷的——cruelty 残忍，残酷
9. -ure 表示：行为，行为的结果，状态，情况
press 压，按——pressure 压力
fail 失败——failure 失败
10. -y 表示：性质；状态；行为，构成抽象名词及学术名
difficult 困难的——difficulty 困难
discover 发明，发现——discovery 发明，发现
possible 可能的——possibility 可能性

Career Training

询 问 观 点

每个人对事情都有自己的观点，因此经常要表达自己的观点。我们表达观点的目的是为了争取别人的理解和赞同，因此在讲话时要注意语气，不要过分生硬，以免令人产生反感情绪。

询问他人观点的常用句型有“What do you think of it?”“What's your opinion on it?”等。发表自己的观点的常用句型有“From my point of view, …”“As far as I am concerned, …”等。他人表达完观点后，你可以同意，也可以不同意他的观点，同意一般可以说“I entirely agree.”或“I can't agree more”；若不同意，简单一点就可以说“I don't think so.”。

在表达观点时，要避免过于武断，例如，涉及对一个人的看法时，我们在国内往往会听到人们冲口而出说“这个人真差劲”；但在英语里，人们却不会这么表达，充其量说“I don't think much of him.”。

总之，英语在表达态度和观点方面要委婉得多，而且显得比较客观、具体。特别是英国人，他们说话口气温和，很忌讳权威式的口吻和语言。即使是在他们为自己辩护时，也会很注意避免使用第一人称“我”。例如，英国人通常会说“Surely, you'd admit that the problem is serious.”；而我们中国人常说“I think the problem is serious.”。

软件水平考试试题解析

【真题再现】

从供选择的答案中选出应填入下列英语文句中____内的正确答案，把编号写在答卷的对应栏内。

The computer itself does not do all the work on its own. The work is done by a __A__ of the computer, called hardware, and __B__ of instructions, called software or computer programs. Inside the machine, the instructions are __C__ and carried out to do the work you want to do. A computer without software is nothing more than a mass of metal and plastic. On the other hand, software without a computer is simply wasted ___D___ because only the computer can use the software and put it to work. When you talk about a computer's being able to do this or that, you are really referring to the ___E___ that accepts your commands and the computer that carries them out.

供选择的答案：

A：①memory	②chip	③combination	④wire
B：①programs	②procedures	③sets	④subroutines
C：①composed	②explained	③interpreted	④organized
D：①disk	②file	③tape	④potential
E：①editor	②hardware	③keyboard	④software

【答案】A：③　B：③　C：③　D：④　E：④

【试题解析】

A：memory 的意思是"存储器，内存"，chip 的意思是"芯片"，combination 的意思是"结合；联合"，wire 的意思是"电线"。该句的意思是工作是由计算机机体（被称作硬件）和指令集（被称作软件或计算机程序）共同来完成的。故选③。

B：set of instructions 是指令集，故选③。

C：输入指令后，计算机先翻译后执行。故选③。

D：disk 的意思是"磁盘"，file 的意思是"文件"，tape 的意思是"录音带，磁带"，potential 的意思是"潜能，潜力"。软件不是磁盘，也不是录音带、磁带。wasted potential 的意思是"被浪费的潜能"，即废物。本句意为：没有计算机，软件也只是废物。故选④。

E：editor 的意思是"编辑器"，hardware 的意思是"硬件"，keyboard 的意思是"键盘"，software 的意思是"软件"。软件接受指令，硬件执行指合。故选④。

【参考译文】

计算机本身并不能做任何工作，工作是由计算机机体（被称作硬件）和指令集（被称作软件或计算机程序）共同来完成的。在计算机内，指令被翻译并执行，以满足工作的要求。如果没有软件，一台计算机不过是一堆金属和塑料。没有计算机，软件也只是废物，因为只有计算机才能调用软件并使其工作。在谈论到计算机能为你做这做那时，实际上是指软件接受你的指令，再由计算机去执行这些指令。

Exercises

[Ex. 1] 根据 Text A 回答以下问题。

1) What is big data?

2) What is a primary goal for looking at big data?

3) Why is big data analytics often associated with cloud computing?

4) What is big data analytics?

5) What is the goal of big data management?

6) What is the difference between textual unstructured data and non-textual unstructured data?

7) What is data mining generally? What is it called sometimes?

8) What parameters do data mining include?

9) What is a data warehouse? Where is it housed?

10) What do applications of data warehouses include?

[Ex. 2] 根据 **Text B** 回答以下问题。

1) What is data mining?

2) Where does data mining derive its name from?

3) What does data mining help analysts do？And how?

4) What are the specific uses of data mining mentioned in the passage?

5) What is a typical example of a predictive problem? What is an example of pattern discovery?

6) Why are financial analysts plowing through vast sets of financial records, data feeds, and other information sources? Why are health-care organizations examining medical records?

7) What is modeling?

8) What are some of the tools used for data mining?

9) What is the result of Merck-Medco's data mining project?

10) What is the future of data mining?

[Ex. 3] 把下列句子翻译为中文。

1) It is anticipated that such cloud computing would accelerate the emergence of"big data".

2) Now most organizations cannot use such unstructured data efficiently.

3) Big data allow organizations to create ever-narrower segmentations and to tailor services precisely to meet customer needs.

4) Much of unstructured data such as documents, images, or audio could not be easily addressed within this scenario.

5) What do you do if the amount of structured, semi-structured, and unstructured data is roughly equal?

6) According to a survey, about 30% of the companies use big data analytics to help them making decisions.

7) Data mining, at its core, is the transformation of large amount of data into meaningful patterns and rules.

8) However, for the average user, clustering can be the most useful data mining method you can use.

9) A data warehouse should be designed for data analysis.

10) The end users might not be the decision makers, but they provide valuable information about the usability of the data warehouse.

[Ex. 4] 软件水平考试真题自测。

从供选择的答案中选出应填入下列英语文句中____内的正确答案，把编号写在答卷的对应栏内。

Applications put computers to practical business __A__ , but below the __B__ it's the heart of an operating system—the kernel—that provides the technical wizardry to juggle multiple program, connect to networks and store __C__ .

A traditional kernel provides all the functions for applications. The kernel __D__ memory, I/O devices and parcels out processor time.

The kernel also supports security and fault __E__ , which is the ability to recover automatically when parts of the system fail.

供选择的答案：

A：①used ②use ③apply ④applied

B：①earth ②bottom ③table ④surface

C：①graphics ②data ③text ④image

D：①manages ②manage ③managed ④managing

E：①error ②question ③tolerance ④problem

[Ex. 5] 听短文填空。

Exercises 10

Business intelligence (BI) comprises the strategies and technologies used by enterprises for the data analysis of business information. BI technologies provide historical, ____1____ and predictive views of business operations. Common functions of BI technologies include ____2____, online analytical processing, analytics, data ____3____, process mining, complex event processing, business performance management, benchmarking, text mining, predictive analytics and prescriptive analytics. BI technologies can handle large amounts of ____4____ and sometimes unstructured data to help identify, develop and otherwise create new strategic business opportunities. They aim to allow for the easy interpretation of these big data. Identifying new opportunities and implementing an effective ____5____ based on insights can provide businesses with a competitive market advantage and long-term stability.

BI can be used by enterprises to support a wide range of ____6____ decisions ranging from operational to strategic. Basic operating decisions include product positioning or ____7____. Strategic business decisions involve priorities, goals and directions at the broadest level. In all cases, BI is most effective when it combines data derived from the market in which a company operates (external data) with data from company sources internal to the business such as ____8____ and operations data (internal data). When combined, external and internal data can provide a complete picture which, in effect, creates an"intelligence"that cannot be derived from any singular set of data. Among myriad uses, BI tools empower organizations to gain ____9____ into new markets, to assess demand and suitability of products and services for different ____10____ segments and to gauge the impact of marketing efforts.

Reference Translation

大 数 据

1. 定义

大数据是用来描述公司产生的浩繁的非结构化和半结构化数据的一个通用术语——要把这些数据加载到关系型数据库来分析会耗费大量时间和大量资金。虽然大数据并没有涉及任何具体数量，通常在谈论千万亿字节（2^{50}bytes）和艾字节（2^{60}bytes）时常使用该术语。

观察大数据的一个主要目标是发现可重复的业务模式。人们普遍承认，非结构化数据，其中大部分在文本文件中，至少占一个组织中数据的80%。如果不加管理，企业每一年产生的全部非结构化数据会花费巨额的存储费用。如果审计或诉讼时不能找到信息，不加管理的非结构化数据也可能会带来法律责任。

大数据分析往往与云计算相关，因为实时分析大型数据集需要像 MapReduce（映射-化简）这样的技术框架来将任务分布到数十台、几百台甚至上千台的计算机上。

2. 大数据分析

大数据分析研究大量的多种类型的数据，以揭示隐藏的模式、未知的关系及其他有用的信息。这些信息可以提供有竞争力的优势，以超过对手组织，产生商业利益，例如，更有效的营销和增加收入。

大数据分析的主要目标是，让数据科学家和其他用户分析数量巨大的业务数据以及可能没有被传统的商务智能（BI）程序利用的其他数据源来帮助企业做出更好的业务决策。这些其他数据源可能包括 Web 服务器日志和互联网点击流数据、社交媒体活动报告、移动电话的呼叫详细记录和传感器捕获的信息。有些人专门对这些非结构化数据进行大数据分析，而 Gartner 和 Forrester 等咨询公司则把业务数据和其他结构化数据当作有效的大数据。

可以用软件工具进行大数据分析。这些工具通常作为高级分析学科的一部分，如预测分析和数据挖掘。但用于大数据分析的非结构化数据源可能不适用于传统的数据仓库。此外，传统的数据仓库可能无法满足大数据所带来的需求。因此，一类新的大数据技术已经出现，并正在很多大数据分析环境中使用。与大数据分析相关的技术包括 NoSQL 数据库、Hadoop 和 MapReduce。这些技术构成了支持集群系统中大数据集处理的开源软件框架的核心。

组织在大数据分析项目上可能遇到的困难包括：内部缺乏分析技能和聘用经验丰富分析专家的高额成本；虽然厂商开始提供数据分析技术之间的软件接口，但把 Hadoop 系统与数据仓库加以整合也颇具挑战性。

3. 大数据管理

大数据管理是对大量结构化和非结构化数据的组织、管理和治理。

大数据管理的目标是确保高水平的数据质量、商务智能的可用性和大数据分析应用。企业、政府机构和其他组织采用大数据管理策略，以帮助它们与快速增长的数据池相抗衡，通常涉及千兆字节级甚至千万亿字节级的用不同文件格式保存的信息。有效的大数据管理帮助企业从各种非结构化和半结构化的数据集找到有价值的信息，数据集包括通话详细记录、系统日志和社交媒体网站。

大部分大数据环境超越了关系型数据库和传统的数据仓库平台，融入了处理和存储非传统数据的技术。对收集和分析大数据的日益注重正在形成一些新平台，这些平台把传统数据仓库与用逻辑数据仓库构建的大数据系统相结合。作为这一进程的一部分，这些平台必须决定哪些数据合格、哪些数据可以处理、哪些数据应该被保存和分析，以改善目前的业务流程或提供一个具有竞争优势的业务。这个过程需要细致的数据分类，以便最终能够对较小的数据集进行快速而高效的分析。

4. 大数据即服务（BDaaS）

大数据即服务（BDaaS）是由外部供应商提供统计分析工具或信息，以帮助企业理解并使用从大型信息集中获得的数据，从而获得竞争优势。

鉴于定期产生的非结构化数据数量巨大，BDaaS 利用外部供应商的预测分析技能来管理和评估大型数据集以释放组织资源，而不是通过内部员工来实现这类功能。它可以通过采用数据处理软件或与数据科学家签订服务合同的形式来实现。

BDaaS 是管理服务的一种形式，类似于软件即服务或基础设施即服务。它往往依赖云存储，为这些信息的拥有者和使用信息的供应者维持连续的数据访问。

5. 非结构化数据

非结构化数据是描述任何不在数据库中的企业信息的通用标签。非结构化数据可以是文本的或非文本的。非结构化文本数据产生于电子邮件、PowerPoint 演示文稿、Word 文档、协作软件和即时消息媒体。非结构化非文本数据产生于 JPEG 图像、MP3 音频文件和 Flash 视频文件等媒体。

非结构化数据中包含的信息不总是很容易找到。它要求可以扫描电子和硬拷贝文档中及其他媒体中的数据，以便可以在特定上下文中按照特定的词语搜索并得到相关概念。这就是所谓的语义搜索，也被称为企业搜索。

在面向客户的业务中，可以对包含在非结构化数据中的信息进行分析，以改善客户关系管理和关系营销。随着 Twitter 和 Facebook 这类社交媒体的应用成为主流，可以预期非结构化数据的增长将远远超过结构化数据的增长。

6. 数据挖掘

通常，数据挖掘（有时也称为数据发现或知识发现）是从不同角度分析数据，并总结成有用信息的过程。此类信息可以增加收入、降低成本或两者兼而有之。数据挖掘软件是众多用于数据分析的工具之一。它允许用户从许多不同层面或角度分析数据，对数据进行分类并总结出确定的关系。从技术上说，数据挖掘是从几十个大型关系数据库中寻找关系或模式的过程。

数据挖掘的范围包括：

关联分析——找出一个事件与另一个事件关联的模式。

序列或路径分析——寻找一个事件导致另一个事件的模式。

分类——寻找新的模式（可能会导致数据组织方式的改变，但没关系）。

聚类——在文档群中寻找以前不知道的事实并视觉化呈现。

预测——找出数据中可以合理预测未来的模式（数据挖掘的这个部分被称为预测性分析）。

数据挖掘技术被用在许多研究领域，包括数学、控制论、遗传学和营销学。Web 挖掘是在客户关系管理（CRM）中使用的一种数据挖掘技术，优势在于从网站的大量信息中找出用户的行为模式。

7. 数据仓库

数据仓库是一个企业的中央存储库，该库中包括了企业各个业务系统收集的全部或重要数据。该术语由 W. H. Inmon 提出。IBM 公司有时使用术语“信息仓库”。

通常，数据仓库放在企业的一个大型服务器上。来自不同的联机事务处理（OLTP）应用程序和其他数据源的数据，被选择性地提取并组织到数据仓库的数据库中，供分析应用程序和用户查询用。数据仓库强调采集不同来源的数据，以便进行有用的分析和存取，但一般不从最终用户或需要访问专门的（有时是本地的）数据库的知识工作者的观点出发。后一想法被称为数据集市。

数据仓库的应用包括数据挖掘、Web 挖掘和决策支持系统（DSS）。

Unit 11

Text A

Text A
How E-commerce
Works

How E-commerce Works

Unless you have been living under a rock for the last few years, you have probably heard about e-commerce.[1] And you have heard about it from several different angles. You may have:

- Heard about all of the companies that offer e-commerce because you have been bombarded by their TV and radio ads.
- Read all of the news stories about the shift to e-commerce and the hype that has developed around e-commerce companies.
- Seen the huge valuations that Web companies get in the stock market, even when they don't make a profit.
- Purchased something on the Web, so you have direct personal experience with e-commerce.

Still, you may feel like you don't understand e-commerce at all. What is all the hype about? Why the huge valuations? And most importantly, is there a way for you to participate? If you have an e-commerce idea, how might you get started implementing it? If you have had questions like these, then this article will help out by exposing you to the entire e-commerce space.

1. Elements to conduct e-commerce

You must have the following elements to conduct e-commerce:

- A product.
- A place to sell the product—in e-commerce, a Web site displays the products in some way and acts as the place.
- A way to get people to come to your Web site.
- A way to accept orders—normally an on-line form of some sort.
- A way to accept money—normally a merchant account handling credit card payments. This piece requires a secure ordering page and a connection to a bank. Or you may use more traditional billing techniques either offline or through the mail.
- A fulfillment facility to ship products to customers (often outsourceable). In the case of software and information, however, fulfillment can occur over the Web through a file download mechanism.
- A way to accept returns.
- A way to handle warranty claims if necessary.
- A way to provide customer service (often through E-mail, on-line forms, on-line knowledge

bases and FAQs, et cetera).

In addition, there is often a strong desire to integrate other business functions or practices into the e-commerce offering. An extremely simple example—you might want to be able to show the customer the exact status of an order.

2. Why the hype?

There is a huge amount of hype that surrounds e-commerce. Given the similarities with mail order commerce, you may be wondering why the hype is so common.[2] Take, for example, the following quotes:

- "On the retail side alone, Forrester projects $17 billion in sales to consumers over the Internet by the year 2001. Some segments are really starting to take off."—Forrester Research.
- "Worldwide business access to the Web is expected to grow at an even faster rate than the US market—from 1.3 million in 1996 to 8 million by 2001."—O'Reilly & Associates.
- "Home continues to be the most popular access location, with nearly 70% of users accessing from their homes...almost 60% shop online. The most popular activities include finding information about a product's price or features, checking on product selection and determining where to purchase a product."—IntelliQuest Information Group, Inc.
- "In general, the more difficult and time-consuming a purchase category is, the more likely consumers will prefer to use the Internet versus standard physical means."—eMarketer.[3]

This sort of hype applies to a wide range of products. According to eMarketer the biggest product categories include:

- Computer products (hardware, software, accessories).
- Books.
- Music.
- Financial Services.
- Entertainment.
- Home Electronics.
- Apparel.
- Gifts and flowers.
- Travel services.
- Toys.
- Tickets.
- Information.

3. Easy and hard aspects of e-commerce

The things that are hard about e-commerce include:

- Getting traffic to come to your Web site.
- Getting traffic to return to your Web site a second time.
- Differentiating yourself from the competition.
- Getting people to buy something from your Web site. Having people look at your site is one thing. Getting them to actually type in their credit card numbers is another.

- Integrating an e-commerce Web site with existing business data (if applicable).

There are so many Web sites, and it is so easy to create a new e-commerce Web site that getting people to look at yours is the biggest problem.

The things that are easy about e-commerce, especially for small businesses and individuals, include:

- Creating the Web site.
- Taking the orders.
- Accepting payment.

There are innumerable companies that will help you build and put up your electronic store.

4. Building an e-commerce site

The things you need to keep in mind when thinking about building an e-commerce site include:[4]

- Suppliers—this is no different from the concern that any normal store or mail order company has. Without good suppliers you cannot offer products.
- Your price point—a big part of e-commerce is the fact that price comparisons are extremely easy for the consumer. Your price point is important in a transparent market.
- Customer relations—e-commerce offers a variety of different ways to relate to your customer. E-mail, FAQs, knowledge bases, forums, chat rooms... Integrating these features into your e-commerce offering helps you differentiate yourself from the competition.
- The back end: fulfillment, returns, customer service—these processes make or break any retail establishment. They define, in a big way, your relationship with your customer.

When you think about e-commerce, you may also want to consider these other desirable capabilities:

- Gift-sending.
- Affiliate programs.
- Special Discounts.
- Repeat buyer programs.
- Seasonal or periodic sales.

The reason why you want to keep these things in mind is because they are all difficult unless your e-commerce software supports them.[5] If the software does support them, they are trivial.

New Words

e-commerce	[i:'kɒmɜ:s]	*n.*电子商务
bombard	[bɒm'bɑ:d]	*vt.*炮轰，轰击
shift	[ʃɪft]	*n.& v.*移动，移位，变化
hype	[haɪp]	*n.*大肆宣传，大做广告
valuation	[ˌvælju'eɪʃn]	*n.*估价，评价，计算
profit	['prɒfɪt]	*n.*利润，益处，得益 *vi.*得益，利用 *vt.*有益于，有利于

participate	[pɑːˈtɪsɪpeɪt]	*vi.*参与，参加，分享，分担
element	[ˈelɪmənt]	*n.*要素，元素，成分，元件
product	[ˈprɒdʌkt]	*n.*产品，产物，乘积
on-line	[ɒnˈlaɪn]	*n.*联机，在线式
merchant	[ˈmɜːtʃənt]	*n.*商人，批发商，贸易商，店主
		*adj.*商业的，商人的
payment	[ˈpeɪmənt]	*n.*付款，支付
bill	[bɪl]	*n.*钞票，账单，票据，清单
facility	[fəˈsɪlətɪ]	*n.*设备，工具
ship	[ʃɪp]	*v.*运输，载运
outsourceable	[ˈaʊtsɔːsəbl]	*adj.*可外界供应的，可外包的
mechanism	[ˈmekənɪzəm]	*n.*机制
claim	[kleɪm]	*vt.*（根据权利）要求，索赔
etcetera	[ˌetˈsetərə]	*n.*其他，等等
integrate	[ˈɪntɪgreɪt]	*vt.*使成整体，使一体化，求……的积分
accessory	[əkˈsesərɪ]	*n.*附件，零件，附加物
		*adj.*附属的，补充的，副的
entertainment	[ˌentəˈteɪnmənt]	*n.*娱乐
apparel	[əˈpærəl]	*n.*衣服，装饰
gift	[gɪft]	*n.*赠品，礼物
competition	[ˌkɒmpəˈtɪʃn]	*n.*竞争者；竞争，竞赛
supplier	[səˈplaɪə]	*n.*供应商，厂商，供给者
extremely	[ɪkˈstriːmlɪ]	*adv.*极端地，非常地
forum	[ˈfɔːrəm]	*n.*论坛
retail	[ˈriːteɪl]	*n.*零售
		*adj.*零售的
		*v.*零售
establishment	[ɪˈstæblɪʃmənt]	*n.*设施，公司，确立，制定
desirable	[dɪˈzaɪərəbl]	*adj.*值得要的，合意的，称心如意的
affiliate	[əˈfɪlieɪt]	*v.*（使……）加入，被接受为会员
discount	[ˈdɪskaʊnt]	*n.*折扣
seasonal	[ˈsiːzənl]	*adj.*季节性的
trivial	[ˈtrɪvɪəl]	*adj.*普通的，平凡的，价值不高的，微不足道的

Phrases

stock market	股市
personal experience	个人体验
credit card	信用卡
take off	拿掉，取消

put up　　提供，进行
chat room　　聊天室
back end　　后端

Abbreviations

FAQ (Frequently Asked Question)　　常见问题

Notes

[1] Unless you have been living under a rock for the last few years, you have probably heard about e-commerce.

本句中，Unless you have been living under a rock for the last few years 是一个条件状语从句，可以改写为 If you haven't been living under a rock for the last few years。Unless 的意思是"除非"，等于"if not"。请看下例：

We will go and repair our printer unless it rains tomorrow.

如果明天不下雨，我们就去修理打印机。

[2] Given the similarities with mail order commerce, you may be wondering why the hype is so common.

本句中，Given the similarities with mail order commerce 是一个条件状语，Given 是一个介词，意思是"考虑到，鉴于"，类似于 taking (sth.) into account。请看下例：

Given his interest in computer design, I'm sure this job is the right career for him.

考虑到他喜欢计算机设计，我可以肯定这个工作是最适合他的职业。

Given his age, he is a remarkable manager.

鉴于他的年龄，他是一个出色的经理。

[3] "In general, the more difficult and time-consuming a purchase category is, the more likely consumers will prefer to use the Internet versus standard physical means."—eMarketer.

本句中，the more… the more…是一个句型，意思是"越……越……"，the more… the less…的意思是"越……越不……"。请看下例：

The more you practice on computer, the more skillful you become.

计算机的操作越练越熟练。

The more difficult the questions are, the less likely I am to be able to answer them.

问题越困难，我就越不可能回答。

[4] The things you need to keep in mind when thinking about building an e-commerce site include:

本句中，you need to keep in mind when thinking about building an e-commerce site 是一个定语从句，修饰和限定 The things。在该定语从句中，when thinking about building an e-commerce site 作时间状语。

[5] The reason why you want to keep these things in mind is because they are all difficult unless your e-commerce software supports them.

本句中，The reason why you want to keep these things in mind 作主语，why you want to keep these things in mind 是一个定语从句，修饰和限定 The reason。is 是系动词，because they are all difficult unless your e-commerce software supports them 是一个表语从句。本句是系表结构作谓语。在该表语从句中，unless your e-commerce software supports them 是一个条件状语从句，修饰表语从句的谓语 are all difficult。

Text B

Text B
Virtual Reality

Virtual Reality

Unlike real reality (the actual world in which we live), virtual reality means simulating bits of our world (or completely imaginary worlds) using high-performance computers and sensory equipment, like headsets and gloves (Figure 11-1).

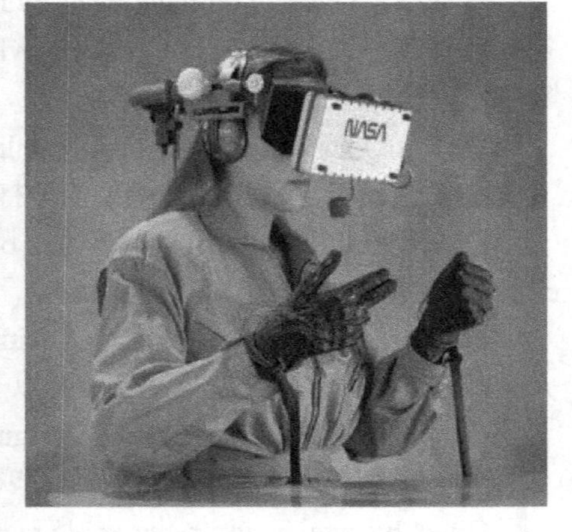

Figure 11-1 Virtual reality device

1. What is virtual reality?

Virtual reality (VR) means experiencing things through our computers that don't really exist. From that simple definition, the idea doesn't sound especially new. When you look at an amazing Canaletto painting, for example, you're experiencing the sites and sounds of Italy as it was about 250 years ago — so that's a kind of virtual reality. In the same way, if you listen to ambient instrumental or classical music with your eyes closed, and start dreaming about things, isn't that an example of virtual reality — an experience of a world that doesn't really exist? What about losing yourself in a book or a movie? Surely that's a kind of virtual reality.

If we're going to understand why books, movies, paintings, and pieces of music aren't the same thing as virtual reality, we need to define VR fairly clearly. For this purpose, I'm going to define it as: a believable, interactive 3D computer-created world that you can explore so you feel you really are there, both mentally and physically.

Putting it another way, virtual reality is essentially:

Believable—you really need to feel like you're in your virtual world (on Mars, or wherever) and to keep believing that, or the illusion of virtual reality will disappear.

Interactive—as you move around, the VR world needs to move with you. You can watch a 3D movie and be transported up to the Moon or down to the seabed — but it's not interactive in any sense.

Computer-generated—why is that important? Because only powerful machines with realistic 3D computer graphics are fast enough to make believable, interactive, alternative worlds that change in real-time as we move around them.

Explorable—a VR world needs to be big and detailed enough for you to explore. However realistic a painting is, it shows only one scene from one perspective. A book can describe a vast and complex "virtual world," but you can only really explore it in a linear way, exactly as the author describes it.

Immersive—to be both believable and interactive, VR needs to engage both your body and your mind. Paintings by war artists can give us glimpses of conflict, but they can never fully convey the sight, sound, smell, taste, and feel of battles. You can play a flight simulator game on your home PC and be lost in a very realistic, interactive experience for hours (the landscape will constantly change as your plane flies through it), but it's not like using a real flight simulator (where you sit in a hydraulically operated mockup of a real cockpit and feel actual forces as it tips and tilts), and even less like flying a plane.

2. Types of virtual reality

"Virtual reality" has often been used as a marketing buzzword for compelling, interactive video games or even 3D movies and television programs, none of which really count as VR because they don't immerse you either fully or partially in a virtual world. Search for "virtual reality" in your cellphone app store and you'll find hundreds of hits, even though a tiny cellphone screen could never get anywhere near producing the convincing experience of VR. Nevertheless, things like interactive games and computer simulations would certainly meet parts of our definition up above, so there's clearly more than one approach to building virtual worlds — and more than one flavor of virtual reality. Here are a few of the bigger variations:

2.1 Fully immersive

For the complete VR experience, we need three things. First, a plausible and richly detailed virtual world to explore; a computer model or simulation, in other words. Second, a powerful computer that can detect what we're doing and adjust our experience accordingly, in real time (so what we see or hear changes as fast as we move—just like in real reality). Third, hardware linked to the computer that fully immerses us in the virtual world as we roam around. Usually, we'd need to put on what's called a head mounted display (HMD) with two screens and stereo sound, and wear one or more sensory gloves. Alternatively, we could move around inside a room, fitted out with surround-sound loudspeakers, onto which changing images are projected from outside. We'll explore VR equipment in more detail in a moment.

2.2 Nonimmersive

A highly realistic flight simulator on a home PC might qualify as nonimmersive virtual reality, especially if it uses a very wide screen, with headphones or surround sound, and a realistic joystick and other controls. Not everyone wants or needs to be fully immersed in an alternative reality. An architect might build a detailed 3D model of a new building to show to clients that can be explored on a desktop computer by moving a mouse. Most people would classify that as a kind of virtual reality, even if it doesn't fully immerse you. In the same way, computer archaeologists often create engaging 3D reconstructions of long-lost settlements that you can move around and explore. They don't take you back hundreds or thousands of years or create the sounds, smells, and tastes of prehistory, but they give a much richer experience than a few pastel drawings or even an animated movie.

2.3 Collaborative

What about "virtual world" games like Second Life and Minecraft? Do they count as virtual reality? Although they meet the first four of our criteria (believable, interactive, computer-created and explorable), they don't really meet the fifth: they don't fully immerse you. But one thing they do offer that cutting-edge VR typically doesn't is collaboration: the idea of sharing an experience in a virtual world with other people, often in real time or something very close to it. Collaboration and sharing are likely to become increasingly important features of VR in future.

2.4 Web-based

Virtual reality was one of the hottest, fastest-growing technologies in the late 1980s and early 1990s, but the rapid rise of the World Wide Web largely killed off interest after that. Even though computer scientists developed a way of building virtual worlds on the Web (using a technology analogous to HTML called Virtual Reality Markup Language, VRML), ordinary people were much more interested in the way the Web gave them new ways to access real reality — new ways to find and publish information, shop, and share thoughts, ideas, and experiences with friends through social media. With Facebook's growing interest in the technology, the future of VR seems likely to be both Web-based and collaborative.

2.5 Augmented reality

Mobile devices like smartphones and tablets have put what used to be supercomputer power in our hands and pockets. If we're wandering round the world, maybe visiting a heritage site like the pyramids or a fascinating foreign city we've never been to before, what we want is typically not virtual reality but an enhanced experience of the exciting reality we can see in front of us. That's spawned the idea of augmented reality (AR), where, for example, you point your smartphone at a landmark or a striking building and interesting information about it pops up automatically. Augmented reality is all about connecting the real world we experience to the vast virtual world of information that we've collectively created on the Web. Neither of these worlds is virtual, but the idea of exploring and navigating the two simultaneously does, nevertheless, have things in common with virtual reality. For example, how can a mobile device figure out its precise location in the world? How do the things you see on the screen of your tablet change as you wander round a city? Technically, these problems are similar to the ones developers of VR systems have to solve — so there are close links between AR and VR.

3. What equipment do we need for virtual reality?

Close your eyes and think of virtual reality and you probably picture something like our top photo: a geek wearing a wraparound headset (HMD) and data gloves, wired into a powerful workstation or supercomputer. What differentiates VR from an ordinary computer experience (using your PC to write an essay or play games) is the nature of the input and output. Where an ordinary computer uses things like a keyboard, mouse, or speech recognition for input, VR uses sensors that detect how your body is moving. And where a PC displays output on a screen (or a printer), VR uses two screens (one for each eye), stereo or surround-sound speakers, and maybe some forms of haptic (touch and body perception) feedback as well. Let's take a quick tour through some of the more common VR input and output devices.

3.1 Head mounted displays (HMDs)

A head mounted display looks like a giant motorcycle helmet or welding visor, but consists of two small screens (one in front of each eye), a blackout blindfold that blocks out all other light (eliminating distractions from the real world), and stereo headphones. The two screens display slightly different, stereoscopic images, creating a realistic 3D perspective of the virtual world. HMDs usually also have built-in accelerometers or position sensors so they can detect exactly how your head and body are moving (both position and orientation — which way they're tilting or pointing) and adjust the picture accordingly. The trouble with HMDs is that they're quite heavy, so they can be tiring to wear for long periods. Some of the really heavy ones are even mounted on stands with counterweights. But HMDs don't have to be so elaborate and sophisticated. At the opposite end of the spectrum, Google has developed an affordable, low-cost pair of cardboard goggles with built-in lenses that convert an ordinary smartphone into a crude HMD.

3.2 Immersive rooms

An alternative to putting on an HMD is to sit or stand inside a room onto whose walls changing images are projected from outside. As you move in the room, the images change accordingly. Flight simulators use this technique, often with images of landscapes, cities, and airport approaches projected onto large screens positioned just outside a mockup of a cockpit. A famous 1990s VR experiment called CAVE (cave automatic virtual environment), developed at the University of Illinois by Thomas de Fanti, also worked this way. People moved around inside a large cube-shaped room with semi-transparent walls onto which stereo images were back-projected from outside. Although they didn't have to wear HMDs, they did need stereo glasses to experience full 3D perception.

3.3 Data gloves

See something amazing and your natural instinct is to reach out and touch it — even babies do that. So giving people the ability to handle virtual objects has always been a big part of VR. Usually, this is done using data gloves, which are ordinary gloves with sensors wired to the outside to detect hand and figure motions. One technical method of doing this uses fiber-optic cables which stretches the length of each finger. Each cable has tiny cuts in it so, as you flex your fingers back and forth, more or less light escapes. A photocell at the end of the cable measures how much light reaches it and the computer uses this to figure out exactly what your fingers are doing. Other gloves use strain gauges, piezoelectric sensors, or electromechanical devices (such as potentiometers) to measure finger movements.

3.4 Wands

Even simpler than a data glove, a wand is a stick you can use to touch, point to, or otherwise interact with a virtual world. It has position or motion sensors (such as accelerometers) built in, along with mouse-like buttons or scroll wheels. Originally, wands were clumsily wired into the main VR computer; increasingly, they're wireless.

New Words

simulating	['sɪmjʊleɪtɪŋ]	*n*.模拟
		v.模仿；（用计算机或模型等）模拟
imaginary	[ɪ'mædʒɪnərɪ]	*adj*.想象中的，假想的，虚构的
sensory	['sensərɪ]	*adj*.传递感觉的

headset ['hedset] *n.*头套

glove [glʌv] *n.*手套

substitute ['sʌbstɪtju:t] *v.*代替，替换，代用

wraparound ['ræpəˌraʊnd] *adj.*包着的，包围的，环绕的

amaze [ə'meɪz] *vt.*使大为吃惊，使惊奇
*n.*吃惊，好奇

ambient ['æmbɪənt] *adj.*周围的，环境的

instrumental [ˌɪnstrə'mentl] *adj.*乐器的；仪器的

fairly ['feəlɪ] *adv.*适当地；清楚地；完全，简直

believable [bɪ'li:vəbl] *adj.*可信的

mentally ['mentəlɪ] *adv.*心理上，精神上；智力上

essentially [ɪ'senʃəlɪ] *adv.*本质上，根本上

illusion [ɪ'lu:ʒn] *n.*错觉；幻想；假象

disappear [ˌdɪsə'pɪə] *vi.*不见，不显示，消失；不复存在

seabed ['si:bed] *n.*海底，海床

realistic [ˌri:ə'lɪstɪk] *adj.*逼真的；栩栩如生的

explorable [ɪk'splɔ:rəbl] *adj.*可探索的

scene [si:n] *n.*场面，现场

linear ['lɪnɪə] *adj.*直线的；线性的

immersive [ɪ'mɜ:sɪv] *adj.*拟真的，沉浸式的，浸入的

glimpse [glɪmps] *n.*一瞥，一看
*vi.*瞥见

convey [kən'veɪ] *vt.*传达，传递；运送，输送；表达

taste [teɪst] *n.*味道，滋味；味觉

simulator ['sɪmjuleɪtə] *n.*模拟装置，模拟器

experience [ɪk'spɪərɪəns] *n.*体验，经验；经历，阅历
*vt.*亲身参与，亲身经历；感受

mockup ['mɒkʌp] *n.*制造模型；制造样机；实体模型

cockpit ['kɒkpɪt] *n.*驾驶员座舱

buzzword ['bʌzwɜ:d] *n.*（报刊等的）时髦术语，流行词

partially ['pɑ:ʃəlɪ] *adv.*部分地

plausible ['plɔ:zəbl] *adj.*貌似真实的

nonimmersive [nɒnɪ'mɜ:sɪv] *adj.*非拟真的，非沉浸式的，非浸入的

joystick ['dʒɔɪstɪk] *n.*操纵杆

archaeologist [ˌɑ:kɪ'ɒlədʒɪst] *n.*考古学家

engage [ɪn'geɪdʒ] *vt.*吸引住
*vi.*与……建立密切关系；衔接；紧密结合

reconstruction [ˌri:kən'strʌkʃn] *n.*复原物；重建物；重建；再现

settlement ['setlmənt] *n.*定居点

prehistory	[ˌpriːˈhɪstrɪ]	*n.*史前时期
cutting-edge	[ˈkʌtɪŋˈedʒ]	*adj.*前沿的
increasingly	[ɪnˈkriːsɪŋlɪ]	*adv.*越来越多地；日益，格外；越来越…
pocket	[ˈpɒkɪt]	*n.*口袋
fascinating	[ˈfæsɪneɪtɪŋ]	*adj.*迷人的，有极大吸引力的
enhance	[ɪnˈhɑːns]	*vt.*提高，增加；加强
spawn	[spɔːn]	*vt.*大量生产
landmark	[ˈlændmɑːk]	*n.*界标；里程碑；纪念碑
simultaneously	[ˌsɪmlˈteɪnɪəslɪ]	*adv.*同时地；一齐
geek	[giːk]	*n.*极客；对电脑痴迷的人
haptic	[ˈhæptɪk]	*adj.*触觉的
feedback	[ˈfiːdbæk]	*n.*反馈；反应
distraction	[dɪˈstrækʃn]	*n.*干扰，注意力分散
stereoscopic	[ˌsterɪəˈskɒpɪk]	*adj.*有立体感的
accelerometer	[əkˌseləˈrɒmɪtə]	*n.*加速度计
orientation	[ˌɔːrɪənˈteɪʃn]	*n.*方向，定位，取向
tilt	[tɪlt]	*v.*倾斜 *n.*倾斜，歪斜
accordingly	[əˈkɔːdɪŋlɪ]	*adv.*相应地；照着，依据
counterweight	[ˈkaʊntəweɪt]	*n.*平衡物，平衡锤，平衡力
elaborate	[ɪˈlæbəreɪt]	*vi.*详尽说明；变得复杂 *vt.*详细制定；尽心竭力地做
sophisticated	[səˈfɪstɪkeɪtɪd]	*adj.*复杂的；精致的；富有经验的
spectrum	[ˈspektrəm]	*n.*光谱；波谱；范围；系列
affordable	[əˈfɔːdəbl]	*adj.*买得起的；价格合理的；负担得起的
lens	[lenz]	*n.*镜头；透镜
crude	[kruːd]	*adj.*粗糙的，简陋的
semi-transparent	[ˈsemɪtrænsˈpærənt]	*adj.*半透明的
wear	[weə]	*vt.*穿着，戴着
fiber-optic	[ˈfaɪbəˈɔptɪk]	*n.*光导纤维
stretch	[stretʃ]	*v.*伸展；延伸
photocell	[ˈfəʊtəʊsel]	*n.*光电池
piezoelectric	[paɪˌiːzəʊɪˈlektrɪk]	*adj.*压电的
potentiometer	[pəˌtenʃɪˈɒmɪtə]	*n.*电位计

Phrases

high-performance computer	高性能计算机
sensory equipment	传感设备
classical music	古典音乐

be transported up to ...	被输送到……
flight simulator	飞行模拟装置，飞行模拟器
television program	电视节目
interactive game	互动游戏
computer model	计算机模型
stereo sound	立体声
surround-sound loudspeaker	环绕声扬声器
a kind of...	……的一种
World Wide Web	万维网
striking building	引人注目的建筑物
pop up	弹出
figure out	弄清；计算出
wraparound headset	环绕式耳机
motorcycle helmet	摩托车头盔
welding visor	焊接面罩
blackout blindfold	遮光眼罩
eliminating distraction	消除干扰
position sensor	位置检测器，位置传感器
cardboard goggles	纸板护目镜
reach out	（使）伸出
piezoelectric sensor	压电传感器
electromechanical device	机电装置，电气机械装置
scroll wheel	滚轮，滚动轮，卷轴轮

Abbreviations

VR (Virtual Reality)	虚拟现实
3D (3 Dimensions)	三维、三个维度、三个坐标
PC (Personal Computer)	个人计算机
HMD (Head Mounted Display)	头戴式显示器
HTML(Hyper Text Markup Language)	超文本标识语言
VRML (Virtual Reality Markup Language)	虚拟现实标识语言
AR (Augment Reality)	增强现实
CAVE (Cave Automatic Virtual Environment)	洞穴式自动虚拟环境

Word Building

常见的构成形容词的后缀列举如下：

1. -able 表示：能……的；可……的；易于……的；

adjust 调整——adjustable 可调整的

suit 适合——suitable 适当的，合适的
2. -al 表示：属于……的；与……有关的
digit 数字——digital 数字的
commerce 商业——commercial 商业的
element 元素——elemental 基本的，元素的
globe 地球——global 全球的，全面的
3. -ed 表示：有……的；如……的
colour 颜色——coloured 有色的，彩色的
gift 才能，天才——gifted 有天才的
skill 技能，技术——skilled 熟练的；有技能的
4. -en 表示：由……制成的；含有……质的；似……的
wool 羊毛——woolen 羊毛制的
wood 木头——wooden 木制的
gold 金子——golden 金制的，似金的
5. -ful 表示：充满……的；具有……性质的；可……的
use 使用——useful 有用的
hope 希望——hopeful 富有希望的
wonder 惊喜，奇迹——wonderful 奇妙的，惊人的
6. -ic 表示：与……有关的；属于……的；有……特性的
atom 原子——atomic 原子的，原子能的
period 时期——periodic 周期的
7. -ive 表示：有……性质的；属于……的；与……有关的；有……倾向的
attract 吸引——attractive 有吸引力的
act 做——active 积极的，主动的
8. -less 表示：没有；无；不
Use 使用——useless 无用的
harm 损害——harmless 无害的
care 小心，仔细——careless 粗心的
wire 线——wireless 无线的
9. -ly 表示：如……的；有……什么性质的
friend 朋友——friendly 友好的
love 爱，喜欢——lovely 可爱的，令人愉快的
time 时间——timely 及时的，适时的
10. -ous 表示：多……的；有……的
danger 危险——dangerous 危险的
poison 毒——poisonous 有毒的
11. -some 表示：充满……的；易于……的；产生……的
tire 使疲惫——tiresome 令人厌倦的

trouble 麻烦——troublesome 令人烦恼的
labour 劳力，劳动——laboursome 费力的，辛苦的
12. -y 表示：多……的；有……的；有点……的
rain 雨——rainy 有雨的
fun 乐趣——funny 有趣的，滑稽可笑的
health 健康——healthy 健康的
noise 噪音——noisy 嘈杂

Career Training

信　函

信函通常由 6 部分构成：抬头、日期、称呼、正文、信尾客套话和签名。在大多数非正式信件中，抬头可以省略。

信函通常用于社会交往，如通报产品信息、协商合作事宜、协调出现问题、进行技术交流等。

School of Computer Science South China University of Technology Guangzhou,510641 China	Heading(抬头)
18 May, 2020	Date(日期)
Dear Sir,	Salutation(称呼)
Body of Letter	Body(正文)
Yours sincerely,	Complimentary Close (信尾客套话)
Wei Feng	Signature(签名)

电子邮件，即 E-mail（electronic mail），具有方便、快捷、低廉的特点，因此得到了广泛使用。

1. E-mail 的基本部分

（1）地址　首先，必须要在标题（Heading）栏的“收件人”（To）框中输入收信人的 E-mail 地址。

（2）主题（Subject）　主题应简明地概括信的内容。它可以是一个单词，也可以是一个名词性短语，或是完整的句子。长度一般不超过 35 个字母。主题的内容切忌含糊不清。一般来说，只要将位于句首的单词和专有名词的首字母大写即可。另外一种较为正规的格式是将少于 5 个字母的介词、连接词或冠词之外的每一个单词的首字母大写，如 New E-mail

Address Notification。

（3）称呼（Salutation） E-mail 一般用于非正式场合，因此正文（Body）前的称呼无须过于严谨。在同辈的亲朋好友或同事之间可以直呼其名，但对长辈或上级最好使用头衔并加上姓。如：Tommy，Mr. Smith。

（4）内容 E-mail 的内容应简明，长篇内容应作为附件。一个段落大多仅由 1～3 个句子组成。

（5）信尾客套话（Complimentary close） 信尾客套话通常也很简明，常常只需一个词，如"Thanks""Best""Cheers"等；不需要用"Sincerely yours"或"Best regards"。

称呼与正文之间、段落之间、正文与信尾客套话之间，一般应空一行。开头无须空格。

电子邮件的典型例子如下：

Jimmy,

I received your memo and will discuss it with Eric on Wednesday.

Best,

David

2. E-mail 的特殊缩略语

在电子邮件的使用中，还流行一些由首字母或读音组成的缩略词。常用的有：

d u wnt 2 go out 2nite：Do you want to go out tonight?

lol：Laughing out loud.

oic：Oh, I see.

mte：My thoughts exactly.

brb：I'll be right back.

c u 2morrow：See you tomorrow.

fanx 4 ur elp：Thanks for your help.

gr8：great.

btw：By the way.

imho：In my humble opinion.

asap：As soon as possible.

另外，网友们为了给枯燥的文字加入感情色彩，还创造了一些表示感情的符号，如 smiley 的符号是：-)，代表"I'm smiling at this"；：-（代表"I'm sad about this"。

上述缩略词和表情符号要视具体情况而定，不宜滥用。

3. 拼写检查

不能以为 E-mail 是非正式文档，就马虎行事。写完之后，一定要认真检查有无拼写、语法和标点符号的错误。可以使用一些软件的拼写检查功能。检查无误后，再发送。特别值得注意的是，对于重要的业务邮件，一定要认真检查拼写。

软件水平考试试题解析

【真题再现】

从供选择的答案中选出应填入下列英语文句中____内的正确答案，把编号写在答卷的对应栏内。

You should be __A__ of developing your program, using something better than the method that uses the philosophy: write __B__ down and then try to get it working. Surprisingly, this method is widely used today with the result that an average programmer on an average job __C__ out only between five to ten lines of correct code per day. We hope your __D__ will be greater. But to improve requires that you apply some discipline to the __E__ of creating programs.

供选择的答案:

A：①available ②capable ③useful ④valuable

B：①anything ②nothing ③something ④thing

C：①does ②looks ③turns ④runs

D：①activity ②code ③productivity ④program

E：①process ②experience ③habit ④idea

【答案】 A：② B：③ C：③ D：③ E：①

【试题解析】

A：be capable of sth./doing sth.的意思是"能够做某事，有做某事的能力"。通常用人或动物作主语，也可以用物作主语。如：He is capable of judging art，即"他具有鉴赏艺术的能力"；This room is capable of 20 people，即"这个房间可容纳 20 人"。available 的意思是"可用到的，可利用的，有用的，有空的"，主语既可以是人也可以是物，人作主语时表示某人有空，物作主语表示可以可得到某物，可利用某物。如：I'm sorry those overcoats are not available in your colour and size，即"对不起，这种外套没有你要的颜色和尺码"；Is the manager available?，即"经理在不在？"。useful 的意思是"有用的, 有益的"。如：That is a useful knife，即"那是把有用的小刀"。valuable 的意思是"贵重的，有价值的，颇有价值的"。如：valuable jewels 意为贵重的珠宝，valuable information 意为有价值的情报。根据句意，选②。

B：something 在这里指代某个指令，nothing 和 anything 不符合逻辑，故选③。

C：look out 的意思是"留神，小心"，turn out 的意思是"生产，制造"，run out 的意思是"用完，耗尽"，do out 的意思是"打扫，收拾"。根据句意，选③。

D：根据前一句话可以知道，平均每位程序员每天只能编写到 5～10 行普通程序代码，作者认为此效率太低，因此希望提高生产率。4 个选项中只有 productivity 符合题意，它的意思是"生产率"，故选③。

E：process 的意思是"过程"，experience 的意思是"经验"，habit 的意思是"习惯，习性"，idea 的意思是"计划；打算；想法"。根据句意，应该是在编写程序时遵循某些确定的原则。故选①。

【参考译文】

你应当能够用比下述方法更有效的方法来开发程序。该方法是把一个指令写下来，然后想方设法使其正常运行。令人吃惊的是，这种方法在今天被大量应用的结果是平均每个程序员每天只能编写到 5～10 行普通程序代码。我们希望您的生产效率能够有所改善，但这要求您在写程序时要遵循某些确定的原则。

Exercises

[Ex. 1] 根据 Text A 回答以下问题。

1) Please list all the elements you must have to conduct e-commerce.

2) What are the things that are easy about e-commerce, especially for small businesses and individuals?

3) What are the things that are hard about e-commerce?

4) What are the things you need to keep in mind when thinking about building an e-commerce site?

5) When you think about e-commerce, what may you also want to consider?

[Ex. 2] 根据 Text B 填空。

1) Virtual reality (VR) means ______________ through our computers that ____________.

2) The first four of criteria of VR are ____________, ____________, ____________ and ____________.

3) However realistic a painting is, it shows only ____________ from ____________.

4) To be both believable and interactive, VR needs to engage both __________ and __________.

5) Paintings by war artists can give us glimpses of conflict, but they can never fully convey __________, __________, __________, __________ and feel of battles.

6) "Virtual reality" has often been used as a marketing buzzword for __________, __________ video games or even __________ and television programs, none of which really count as VR because they don't immerse you either fully or partially ____________.

7) Augmented reality is all about connecting the real world ____________ to the vast virtual world of information ______________ on the Web.

8) A head-mounted display looks like ________________ or welding visor, but consists of ________________ (one in front of each eye), ________________ that blocks out all other light (eliminating distractions from the real world), and ________________.

9) Data gloves are ordinary gloves with ________ wired to the outside to detect ____________.

10) A wand is a stick you can use to ________, __________, or ________________. It has position or motion sensors (such as accelerometers) built in, along with _______________ or ______________.

[Ex. 3] 把下列句子翻译为中文。

1) Have you ever heard about e-commerce? Yes. One of my friends bought something on the Web.

2) Getting people to come to your Web site is one of the elements to conduct e-commerce.

3) Many people are wondering why there is a huge amount of hype that surrounds e-commerce.

4) You can buy computer products, such as hardware, software, accessories, books and music on the Web.

5) Getting traffic to return to your Web site a second time is very hard.

6) Getting people to buy something from your Web site is important when you conduct e-commerce.

7) Our country has been trying to broaden its commerce with other nations.

8) Creating a Web site is easy, but differentiating yourself from the competition is difficult.

9) When thinking about building an e-commerce site, you need to keep customer relations in mind.

10) Your price point is important in a transparent market, because price comparisons are extremely easy for the consumer.

[Ex. 4] 软件水平考试真题自测。

从供选择的答案中选出应填入下列英语文句中____内的正确答案，把编号写在答卷的对应栏内。

An instruction is made up of operations that __A__ the function to be performed and operands that represent the data to be operated on. For example, if an instruction is to perform the operation of __B__ two numbers, it must know __C__ the two numbers are. The processor's job is to __D__ instructions and operands from memory and to perform each operation. Having done that, it signals memory to send it __E__ instruction.

供选择的答案：

A: ①skip ②smile ③smoke ④specify

B: ①add ②added ③adding ④addition

C: ①when ②where ③which ④who

D: ①get ②make ③push ④put

E: ①ant ②last ③next ④second

[Ex. 5] 听短文填空。

Exercises 11

B2C and B2B are two forms of ___1___ transactions. B2C, which stands for business-to-consumer, is a process for selling products directly to ___2___. B2B, which stands for business-to-business, is a process for selling products or services to other businesses. The business systems that support B2B or B2C communications, transactions and ___3___ administration differ in complexity, scope, scale and cost, so it is important that you implement the ___4___ system for your customers.

Consumers buy your products or services for ___5___ use. Business buyers purchase products or services for use in their companies.

In B2C, consumers who buy products from you pay the same price as other consumers. In B2B, price may ___6___ by customer. Customers who agree to place large orders or negotiate special terms ___7___ different prices to other customers. Payment mechanisms also differ.

In B2C transactions, consumers ___8___ products and pay for them at the point of sales using payment mechanisms such as credit or debit cards, checks or cash. B2B transactions ___9___ a more complex business system. Customers select products, place an order and arrange delivery through an agreed logistics ___10___. Customers do not pay at the time of the order, but receive an invoice which they settle within agreed payment terms.

Reference Translation

电子商务如何运作

除非过去几年你一直生活在岩洞中，否则你应该听说过电子商务。你可能从不同的角度听说过它。也许是：

- 听说了全部提供电子商务的公司，因为你已经被它们的电视和收音机广告“轰击”了。
- 阅读了转向电子商务的全部新闻故事及有关电子商务公司发展的大肆宣传。
- 看到了网络公司在股市所获得的巨大价值，即使它们没有获得利润。
- 在网站上购买过一些东西，有直接与电子商务相关的个人体验。

尽管如此，你也许仍然觉得自己一点也不了解电子商务。这些大肆宣传意欲何为？为何有巨大价值？而且，最重要的是，你有参与的方法吗？如果你想开展电子商务，如何开始实现它？如果你有这些问题，那么本文将通过介绍整个电子商务领域，来帮你解决这些问题。

1. 开展电子商务的要素

要开展电子商务，必须有以下要素：

- 产品。
- 销售产品的地方——在电子商务中，一个网站以某种方式展示产品并作为销售场所。
- 让人们来到你的网站的方法。

- 接受订单的手段——通常为某种在线形式。
- 接受货币的途径——通常是处理信用卡支付的商业账号。这一条要求一个安全的订货页面并要与银行联结。或者可以使用更传统的支付方法，要么线下支付，要么通过汇款支付。
- 送货设施（通常外包）。在交付软件或信息产品时，可以通过网站的下载机制来交付。
- 接受退货的途径。
- 如果必要，处理客户索取保修的手段。
- 提供为客户服务的方法（往往通过电子邮件、在线形式、在线知识库及常见问题解答等）。

另外，人们很希望把其他商务运作整合到电子商务中。特别简单的例子是——也许要给客户展现订单的确切状态。

2．为什么要大肆宣传？

有大量的广告围绕电子商务。鉴于电子商务与信函订购商务类似，你也许想知道为何广告如此普遍。例如，有以下广告：

- “就零售来说，到 2001 年，Forrester 公司计划在因特网上花费 170 亿美元，开展对客户的销售。一部分计划已经启动。”得 Forrester 公司研究报告。
- “预期世界范围的网络商务要比美国市场增长得快——从 1996 年的 130 万增长到 2001 年的 800 万。”——O’Reilly & Associates。
- “家庭仍然是最普遍的访问场所，有大约 70%的用户从家中访问……几乎 60%是在线购物。最常见的活动包括寻找产品定价和功能信息，选定产品并决定在哪里购买。”——IntelliQuest Information Group。
- “一般情况下，购买的种类越难，花费时间越多，与常规途径相比，客户就越会选择因特网。”——eMarketer。

这类宣传适用于多种产品。eMarketer 认为最全面的产品分类包括：

- 计算机产品（硬件、软件及相关产品）。
- 书籍。
- 音乐。
- 金融服务。
- 娱乐。
- 家用电器。
- 服装。
- 礼品和鲜花。
- 旅游服务。
- 玩具。
- 票券。
- 信息。

3. 电子商务的难与易

电子商务的困难之处在于：

- 获得网站流量。

- 获得再次访问网站的流量。
- 在竞争中体现特色。
- 让人们从你的网站上购买商品。让人们看你的网站是一回事，让他们实际输入自己的信用卡号是另一回事。
- 整合电子商务网站与现有的商务数据（如果可行的话）。

现在有许多网站，而且建立一个新的电子商务网站非常容易，以至于如何让人们看你的网站成了最大的问题。

电子商务的容易之处，特别是对小企业和个人，在于：

- 建立网站。
- 接受订单。
- 接受支付。

有数不清的公司可以帮助你建立电子商店。

4. 建立电子商务网站

当考虑建立电子商务网站时，需要记住以下事宜：

- 供应商——这与任何普通商店和信函订购公司所关心的事情相同。没有好的供应商，就不能提供好的产品。
- 价位——电子商务的一个主要特点就是客户比较价格变得极其容易。在一个透明的市场中，价位非常重要。
- 客户关系——电子商务提供了与客户建立关系的多种方法。电子邮件、常见问题解答、知识库、论坛、聊天室等功能被整合到电子商务服务中，有助于在竞争中体现特色。
- 后端：配送、退货、客户服务——这些工作可以成全或损害任何零售机构。在很大程度上，它们确立你和客户的关系。

当考虑电子商务时，还要考虑其他优惠：

- 发送赠品。
- 加入会员。
- 特殊折扣。
- 重复购买优惠。
- 季节性或周期性销售优惠。

需要记住这些事情的理由是，除非你的电子商务软件支持它们，否则业务很难开展。如果你的电子商务软件确实支持它们，这些事情就无足轻重了。

Unit 12

Text A

AI

Artificial Intelligence (AI) is the simulation of human intelligence processes by machines, especially computer systems. These processes include learning (the acquisition of information and rules for using the information), reasoning (using rules to reach approximate or definite conclusions) and self-correction. Particular applications of AI include expert systems, speech recognition and machine vision.

AI can be categorized as either weak or strong. Weak AI, also known as narrow AI, is an AI system that is designed and trained for a particular task. Virtual personal assistants, such as Apple's Siri, are a form of weak AI. Strong AI, also known as artificial general intelligence, is an AI system with generalized human cognitive abilities. When presented with an unfamiliar task, a strong AI system is able to find a solution without human intervention.

Because hardware, software and staffing costs for AI can be expensive, many vendors are including AI components in their standard offerings, as well as access to artificial intelligence as a service (AIaaS) platforms. AI as a service allows individuals and companies to experiment with AI for various business purposes and sample multiple platforms before making a commitment. Popular AI cloud offerings include Amazon AI services, IBM Watson Assistant, Microsoft Cognitive Services and Google AI services.

While AI tools present a range of new functionality for businesses, the use of artificial intelligence raises ethical questions. This is because deep learning algorithms, which underpin many of the most advanced AI tools, are only as smart as the data they are given in training.[1] Because a human selects what data should be used for training an AI program, the potential for human bias is inherent and must be monitored closely.

Some industry experts believe that the term artificial intelligence is too closely linked to popular culture, causing the general public to have unrealistic fears about artificial intelligence and improbable expectations about how it will change the workplace and life in general. Researchers and marketers hope augmented intelligence, which has a more neutral connotation, will help people understand that AI will simply improve products and services and will not replace the humans that use them.[2]

1. Types of AI

Arend Hintze, an assistant professor of integrative biology and computer science and engineering at Michigan State University, categorizes AI into four types, from the kind of AI systems that exist today to sentient systems which do not yet exist.[3] His categories are as follows.

Type 1: reactive machines. An example is Deep Blue, the IBM chess program that beat Garry Kasparov in the 1990s. Deep Blue can identify pieces on the chess board and make predictions, but it has no memory and cannot use past experiences to inform future ones. It analyzes possible moves—its own and its opponent—and chooses the most strategic move. Deep Blue and Google's AlphaGO were designed for narrow purposes and cannot easily be applied to another situation.

Type 2: limited memory. These AI systems can use past experiences to inform future decisions. Some of the decision-making functions in self-driving cars are designed this way. Observations inform actions happening in the not-so-distant future, such as a car changing lanes. These observations are not stored permanently.

Type 3: theory of mind. This psychology term refers to the understanding that others have their own beliefs, desires and intentions that impact the decisions they make. This kind of AI does not yet exist.

Type 4: self-awareness. In this category, AI systems have a sense of self, have consciousness. Machines with self-awareness understand their current state and can use the information to infer what others are feeling. This type of AI does not yet exist.

2. Examples of AI technology

AI is incorporated into a variety of different types of technology. Here are some examples.

Automation: what makes a system or process function automatically. For example, robotic process automation (RPA) can be programmed to perform high-volume, repeatable tasks that humans normally performed. RPA is different from IT automation in that it can adapt to changing circumstances.

Machine learning: the science of getting a computer to act without programming. Deep learning is a subset of machine learning that, in very simple terms, can be thought of as the automation of predictive analytics. There are three types of machine learning algorithms:

- Supervised learning: data sets are labeled so that patterns can be detected and used to label new data sets.
- Unsupervised learning: data sets aren't labeled and are sorted according to similarities or differences.
- Reinforcement learning: data sets aren't labeled but, after performing an action or several actions, the AI system is given feedback.

Machine vision: the science of allowing computers to see. This technology captures and analyzes visual information using a camera, analog-to-digital conversion and digital signal processing. It is often compared to human eyesight, but machine vision isn't bound by biology and can be programmed to see through walls, for example. It is used in a range of applications from signature identification to medical image analysis. Computer vision, which is focused on machine-based image processing, is often conflated with machine vision.

Natural language processing (NLP): the processing of human—and not computer—language by a computer program. One of the older and best known examples of NLP is spam detection, which looks at the subject line and the text of an E-mail and decides if it's junk. Current approaches to NLP are based on machine learning. NLP tasks include text translation, sentiment analysis and speech recognition.

Robotics: a field of engineering focused on the design and manufacturing of robots. Robots are often used to perform tasks that are difficult for humans to perform or perform consistently. They are used in assembly lines for car production or by NASA to move large objects in space. Researchers are also using machine learning to build robots that can interact in social settings.

Self-driving cars: these use a combination of computer vision, image recognition and deep learning to build automated skill at piloting a vehicle while staying in a given lane and avoiding unexpected obstructions, such as pedestrians.

3. AI applications

AI has made its way into a number of areas. Here are six examples.

AI in healthcare. The biggest bets are on improving patient outcomes and reducing costs. Companies are applying machine learning to make better and faster diagnoses than humans. One of the best known healthcare technologies is IBM Watson. It understands natural language and is capable of responding to questions asked of it. The system mines patient data and other available data sources to form a hypothesis, which then presents with a confidence scoring schema. Other AI applications include chatbots, a computer program used online to answer questions and assist customers, to help schedule follow-up appointments or aid patients through the billing process, and virtual health assistants that provide basic medical feedback.[4]

AI in business. Robotic process automation is being applied to highly repetitive tasks normally performed by humans. Machine learning algorithms are being integrated into analytics and CRM platforms to uncover information on how to better serve customers. Chatbots have been incorporated into websites to provide immediate service to customers. Automation of job positions has also become a talking point among academics and IT analysts.

AI in education. AI can automate grading, giving educators more time. AI can assess students and adapt to their needs, helping them work at their own pace. AI tutors can provide additional support to students, ensuring they stay on track. AI could change where and how students learn, perhaps even replacing some teachers.

AI in finance. AI in personal finance applications, such as Mint or Turbo Tax, is making a break in financial institutions. Applications such as these collect personal data and provide financial advice. Other programs, such as IBM Watson, have been applied to the process of buying a home. Today, software performs much of the trading on Wall Street.

AI in law. The discovery process, sifting through documents, in law is often overwhelming for humans. Automating this process is a more efficient use of time. Startups are also building question-and-answer computer assistants that can sift programmed-to-answer questions by examining the taxonomy and ontology associated with a database.

AI in manufacturing. This is an area that has been at the forefront of incorporating robots into the workflow. Industrial robots used to perform single tasks and were separated from human workers, but as the technology advanced that changed.

4. Security and ethical concerns

The application of AI in the realm of self-driving cars raises security as well as ethical concerns. Cars can be hacked, and when an autonomous vehicle is involved in an accident, liability is unclear. Autonomous vehicles may also be put in a position where an accident is unavoidable, forcing the programming to make an ethical decision about how to minimize damage.

Another major concern is the potential for abuse of AI tools. Hackers are starting to use sophisticated machine learning tools to gain access to sensitive systems, complicating the issue of security beyond its current state.

Deep learning-based video and audio generation tools also present bad actors with the tools necessary to create so-called deepfakes, convincingly fabricated videos of public figures saying or doing things that never took place.

5. Regulation of AI technology

Despite these potential risks, there are few regulations governing the use of AI tools, and where laws do exist, they typically pertain to AI only indirectly. For example, Federal Fair Lending Regulations require financial institutions to explain credit decisions to potential customers, which limit the extent to which lenders can use deep learning algorithms, which by their nature are typically opaque.[5] Europe's GDPR puts strict limits on how enterprises can use consumer data, which impedes the training and functionality of many consumer-orientated AI applications.

In 2016, the National Science and Technology Council (NSTC) issued a report examining the potential role governmental regulation might play in AI development, but it did not recommend specific legislation be considered. Since that time the issue has received little attention from lawmakers.

New Words

acquisition	[ˌækwɪˈzɪʃn]	*n.*获得
rule	[ru:l]	*n.*规则，规定；统治，支配
		*v.*控制，支配
reasoning	[ˈri:zənɪŋ]	*n.*推理，论证
		*v.*推理，思考；争辩；说服（reason的ing形式）
		*adj.*推理的
approximate	[əˈprɒksɪmət]	*adj.*极相似的
		*vi.*接近于，近似于
		*vt.*靠近，使接近
definite	[ˈdefɪnət]	*adj.*明确的；一定的；肯定
conclusion	[kənˈklu:ʒn]	*n.*结论；断定，决定；推论
self-correction	[ˌselfkəˈrekʃn]	*n.*自校正；自我纠错；自我改正
particular	[pəˈtɪkjələ]	*adj.*特别的；详细的；独有的
		*n.*特色，特点
vision	[ˈvɪʒn]	*n.*视觉

narrow	['nærəʊ]	*adj.*狭隘的，狭窄的
virtual	['vɜ:tʃuəl]	*adj.*（计算机）虚拟的；实质上的，事实上的
cognitive	['kɒgnətɪv]	*adj.*认知的，认识的
ability	[ə'bɪlətɪ]	*n.*能力，资格；能耐，才能
unfamiliar	[ˌʌnfə'mɪlɪə]	*adj.*不熟悉的；不常见的；陌生的；没有经验的
experiment	[ɪk'sperɪmənt]	*n.*实验，试验；尝试
		*vi.*做实验
commitment	[kə'mɪtmənt]	*n.*承诺，许诺；委任，委托
ethical	['eθɪkl]	*adj.*道德的，伦理的
underpin	[ˌʌndə'pɪn]	*vt.*加固，支撑
unrealistic	[ˌʌnrɪə'lɪstɪk]	*adj.*不切实际的；不现实的；空想的
fear	[fɪə]	*n.*害怕；可能性
		*vt.*害怕；为……忧虑（或担心、焦虑）
		*vi.*害怕；忧虑
expectation	[ˌekspek'teɪʃn]	*n.*期待；预期
neutral	['nju:trəl]	*adj.*中立的
connotation	[ˌkɒnə'teɪʃn]	*n.*内涵，含义
integrative	['ɪntɪgreɪtɪv]	*adj.*综合的，一体化的
sentient	['sentɪənt]	*adj.*有感觉能力的，有知觉力的
reactive	[rɪ'æktɪv]	*adj.*反应的
prediction	[prɪ'dɪkʃn]	*n.*预测，预报；预言
opponent	[ə'pəʊnənt]	*n.*对手
observation	[ˌɒbzə'veɪʃn]	*n.*观察，观察力
psychology	[saɪ'kɒlədʒɪ]	*n.*心理学；心理特点；心理状态
intention	[ɪn'tenʃn]	*n.*意图，目的；意向
self-awareness	[self-ə'weənɪs]	*n.*自我意识
consciousness	['kɒnʃəsnəs]	*n.*意识，观念；知觉
circumstance	['sɜ:kəmstəns]	*n.*环境，境遇
similarity	[ˌsɪmə'lærətɪ]	*n.*相像性，相仿性，类似性
signature	['sɪgnətʃə]	*n.*签名；署名；识别标志
identification	[aɪˌdentɪfɪ'keɪʃn]	*n.*鉴定，识别
detection	[dɪ'tekʃn]	*n.*检查，检测
junk	[dʒʌŋk]	*vt.*丢弃，废弃
		*n.*废品；假货
consistently	[kən'sɪstəntlɪ]	*adv.*一贯地，坚持地
pilot	['paɪlət]	*n.*引航员；向导
		*vt.*驾驶
vehicle	['vi: əkl]	*n.*车辆；交通工具
unexpected	[ˌʌnɪk'spektɪd]	*adj.*意外的；忽然的；突然的

obstruction	[əb'strʌkʃn]	*n.*阻塞，阻碍，受阻
pedestrian	[pə'destrɪən]	*n.*步行者，行人 *adj.*徒步的
healthcare	['helθkeə]	*n.*卫生保健
diagnose	['daɪəgnəuz]	*vt.*诊断；判断 *vi.*做出诊断
hypothesis	[haɪ'pɒθəsɪs]	*n.*假设，假说；前提
chatbot	['tʃætbɒt]	*n.*聊天机器人
appointment	[ə'pɔɪntmənt]	*n.*预约
repetitive	[rɪ'petətɪv]	*adj.*重复的，啰唆的
overwhelming	[ˌəuvə'welmɪŋ]	*adj.*势不可挡的，压倒一切的
taxonomy	[tæk'sɒnəmɪ]	*n.*分类学，分类系统
ontology	[ɒn'tɒlədʒɪ]	*n.*本体，存在；实体论
forefront	['fɔ:frʌnt]	*n.*前列；第一线；活动中心
incorporating	[ɪn'kɔ:pəreɪtɪŋ]	*v.*融合，包含；使混合
realm	[relm]	*n.*领域，范围
accident	['æksɪdənt]	*n.*意外事件；事故
unclear	[ˌʌn'klɪə]	*adj.*不清楚的，不明白的，含糊不清
unavoidable	[ˌʌnə'vɔɪdəbl]	*adj.*不可避免的，不得已的
minimize	['mɪnɪmaɪz]	*vt.*把……减至最低数量‘程度’，最小化
damage	['dæmɪdʒ]	*n.*损害，损毁；赔偿金 *v.*损害，毁坏
deepfake	['di:pfeɪk]	*n.*“换脸术”，一种人物图像合成技术
convincingly	[kən'vɪnsɪŋlɪ]	*adv.*令人信服地，有说服力地
fabricate	['fæbrɪkeɪt]	*vt.*编造，捏造
regulation	[ˌregjʊ'leɪʃn]	*n.*规章，规则 *adj.*规定的
credit	['kredɪt]	*n.*信誉，信用；[金融]贷款 *vt.*相信，信任
opaque	[əʊ'peɪk]	*adj.*不透明的；含糊的 *n.*不透明
strict	[strɪkt]	*adj.*严格的；精确的；绝对的
impede	[ɪm'pi:d]	*vt.*阻碍；妨碍；阻止
lawmaker	['lɔ:meɪkə]	*n.*立法者

Phrases

human intelligence	人类智能
expert system	专家系统
speech recognition	语音识别

machine vision	机器视觉
be categorized as ...	被分类为……
weak AI	弱人工智能
virtual personal assistant	虚拟个人助理
strong AI	强人工智能
artificial general intelligence	通用人工智能
for ... purpose	为了……目的
a range of	一系列，一些，一套
deep learning algorithm	深度学习算法
computer science	计算机科学
sentient system	感觉系统
self-driving car	自动驾驶汽车
not-so-distant future	不远的将来
a sense of ...	一种……感觉
be incorporated into ...	被并入……
predictive analytic	预测分析
supervised learning	监督学习
unsupervised learning	无监督学习
reinforcement learning	强化学习
analog-to-digital conversion	模（拟）数（字）转换
digital signal	数字信号
medical image analysis	医学图像分析
machine-based image processing	基于机器的图像处理
be conflated with ...	与……混为一谈
spam detection	垃圾邮件检测
text translation	文本翻译
sentiment analysis	情感分析，倾向性分析
assembly line	（工厂产品的）装配线，流水线
social setting	社会环境，社会场景，社会情境
image recognition	图像识别
confidence scoring schema	置信评分模式
virtual health assistant	虚拟健康助理
talking point	话题；论题；论据
financial institution	金融机构

Abbreviations

AI (Artificial Intelligence)	人工智能
AIaaS (Artificial Intelligence as a Service)	人工智能即服务
RPA (Robotic Process Automation)	机器人过程自动化

NLP (Natural Language Processing) 自然语言处理
NASA (National Aeronautics and Space Administration) 美国航空航天局
CRM (Customer Relationship Management) 客户关系管理
GDPR (General Data Protection Regulation) 普通数据保护条例
NSTC (National Science and Technology Council) 国家科学技术委员会

Notes

[1] This is because deep learning algorithms, which underpin many of the most advanced AI tools, are only as smart as the data they are given in training.

本句中，because deep learning algorithms, which underpin many of the most advanced AI tools, are only as smart as the data they are given in training是一个表语从句。在该从句中，which underpin many of the most advanced AI tools是一个非限定性定语从句，修饰deep learning algorithms，对其进行补充说明。they are given in training是一个定语从句，修饰和限定the data。as... as...的意思是“和……一样……”。

[2] Researchers and marketers hope augmented intelligence, which has a more neutral connotation, will help people understand that AI will simply improve products and services and will not replace the humans that use them.

本句中，which has a more neutral connotation是一个非限定性定语从句，修饰augmented intelligence，对其进行补充说明。that AI will simply improve products and services and will not replace the humans that use them是由and连接的两个并列句，作understand的宾语；that use them是一个定语从句，修饰和限定the humans。

[3] Arend Hintze, an assistant professor of integrative biology and computer science and engineering at Michigan State University, categorizes AI into four types, from the kind of AI systems that exist today to sentient systems which do not yet exist.

本句中，an assistant professor of integrative biology and computer science and engineering at Michigan State University对Arend Hintze进行补充说明，说明Arend Hintze的身份。that exist today是一个定语从句，修饰和限定AI systems。which do not yet exist也是一个定语从句，修饰和限定sentient systems。

[4] Other AI applications include chatbots, a computer program used online to answer questions and assist customers, to help schedule follow-up appointments or aid patients through the billing process, and virtual health assistants that provide basic medical feedback.

本句中，a computer program used online to answer questions and assist customers, to help schedule follow-up appointments or aid patients through the billing process, and virtual health assistants that provide basic medical feedback是一个名词性短语，对chatbots进行补充说明。在该短语中，used online to answer questions and assist customers, to help schedule follow-up appointments or aid patients through the billing process, and virtual health assistants that provide basic medical feedback是一个过去分词短语，作定语，修饰和限定a computer program。在该短语中to answer questions and assist customers, to help schedule follow-up appointments or aid patients through the billing process, and virtual health assistants作目的状语，修饰used。that provide basic medical feedback 是一个定语从句，修饰和限定 virtual health assistants。

[5] For example, Federal Fair Lending Regulations require financial institutions to explain credit decisions to potential customers, which limit the extent to which lenders can use deep learning algorithms, which by their nature are typically opaque.

本句中，which limit the extent to which lenders can use deep learning algorithms, which by their nature are typically opaque 是一个非限定性定语从句，对 Federal Fair Lending Regulations 进行补充说明。在该从句中，to which lenders can use deep learning algorithms 是一个介词前置的定语从句，修饰和限定 the extent。which by their nature are typically opaque 是一个非限定性定语从句，对 deep learning algorithms 进行补充说明。

Text B

What Is Machine Learning?

Artificial intelligence and machine learning are among the most trending technologies these days. Artificial Intelligence teaches computers to behave like a human, to think, and to give a response like a human, and to perform the actions like humans perform.

1. What is machine learning?

As the name suggests, machine learning means the machine is learning.

This is the technique through which we teach the machines about things. It is a branch of artificial intelligence and I would say it is the foundation of artificial intelligence. Here we train our machines using data. If you take a look into it, you'll see that it is something like data mining. Actually, the concept behind it is that machine learning and data mining are both data oriented. We work on data in both of situations. Actually, in data sciences or big data, we analyze the data and make the statistics out of it and we work on how we can maintain our data, how we can conclude the results and make a summary of it instead of maintaining the complete comprehensive bulk of data. But in machine learning, we teach the machines to make the decisions about things. We teach the machine with different data sets and then we check the machine for some situations and see what kind of results we get from this unknown scenario. We also use this trained model for prediction in new scenarios.

We teach the machine with our historical data, observations, and experiments. And then, we predict with the machine from these learnings and take the responses.

As I already said, machine learning is closely related to data mining and statistics.

Data mining—concerned with analytics of data.

Statistics—concerned with prediction-making/probability.

2. Why do we need machine learning?

In this era, we're using wireless communication, Internet etc. Using social media, or driving cars, or anything we're doing right now, is actually generating the data at the backend. If you're surprised about how our cars are generating the data, remember that every car has a small computer inside which controls your vehicle completely, i.e., when which component needs the current, when the specific component needs to start or switch. In this way, we're generating TBs (terabytes) of data.

But this data is also important to get to the results. Let's take an example and try to understand the concept clearly. Let's suppose a person is living in a town and he goes to a shopping mall and buys something. We have many items of a single product. When he buys something, now we can generate the pattern of the things he has bought. In the same way, we can generate the selling and purchasing patterns of things of different people. Now you might be thinking about a random person who comes and buys something and then he never comes again, but we have the pattern of things there as well. With the help of this pattern we can make a decision about the things people most like and when they come to the mall again. They will see the things they want just at the entrance. This is how we attract the customer with machine learning.

3. How do machines learn?

Actually, machines learn through the patterns of data. Let's start with the data sets of data. The input we give to the machine is called x and the response we get is y. Here we've three types of learning.

- Supervised Learning.
- Unsupervised Learning.
- Reinforcement Learning.

3.1 Supervised learning

In supervised learning, we know about the different cases (inputs) and we know the labels (output) of these cases. And here we already know about the basic truths, so here we just focus on the function (operation) because it is the main and most important thing here(Figure 12-1).

Here we just create the function to get the output of the inputs. And we try to create the function which processes the data and try to give the accurate outputs (y) in most of the scenarios(Figure 12-2).

Figure 12-1 Inputs and output of supervised learning

Figure 12-2 Function of supervised learning

Because we've started with known values for our inputs, we can validate the model and make it even better.

And we teach our machine with different data sets. Now it is the time to check it in unknown cases and generate the value.

Note: Let's suppose you've provided the machine a data set of some kind of data and now you train the model according to this data set. Now the result comes to you from this model on the basis of this knowledge set you've provided. But let's suppose if you delete an existing item in this knowledge set or you update something then you don't expect the results you get according to this new modification you've made in the data set.

Now let's see how supervised learning works.

Let's take an example here of iPhone. Let's suppose different customers purchase the iPhone in different years and its price gradually increases.

Here is the general image for any model. Here we can see how many cases we have here. Let's suppose different customers purchase the iPhone with different prices and we know each year, its price increases. We already know that the thing which is independent comes horizontally in the graph and vertical represents the dependent thing. So prize works vertically in the graph. Now draw a line which touches the maximum points. This straight line is the supervised learning of your model. If the next version of iPhone comes in the next year, your model can take a decision how much the prize it will be. This is how we're predicting the values of the things. Although it is not so accurate but it is approximately near to the things.

3.2 Unsupervised learning

It is quite different from supervised learning. Here we don't know about the labels (output) of different cases. And here, we train the model with patterns by finding similarities. And then these patterns become the cluster(Figure 12-3).

Cluster = Collection of similar patterns of data

And then, this cluster is used to analyze and to process the data.

In unsupervised learning, we really don't know if the output is right or wrong. So here in this scenario, the system recognizes the pattern and tries to calculate the results until we get the nearly right value.

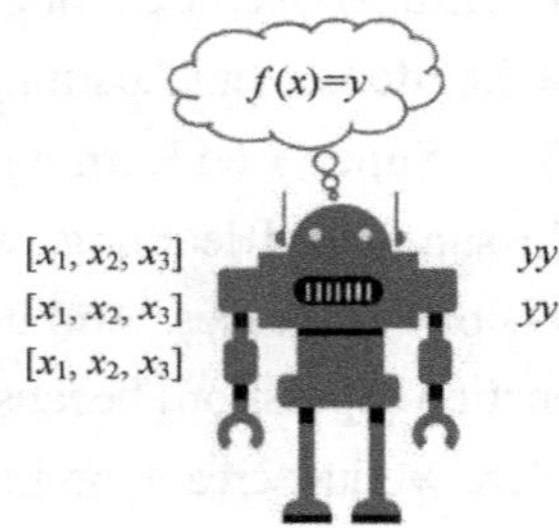

Figure 12-3 Unsupervised learning

3.3 Reinforcement learning

It is like reward based learning. The example of reward based is, suppose your parents will give you a reward on the completion of a specific task. So here you know you've to complete this task and how much it is necessary for you. Here the developer decides himself what reward he'll give on the completion of this task.

It is also feedback oriented learning. Now you're doing some tasks and on the basis of these tasks, you're getting feedback. And if the feedback is positive then it means you're doing it right and you can improve your work on your own. And if the feedback is negative then you know as well what was wrong and how to do it correctly. And feedback comes from the environment where it is working.

It makes the system more optimal than the unsupervised scenario, because here we have some clues like rewards or good feedback to make our system more efficient.

4. Steps in machine learning

There are some key point steps of machine learning when we start to teach the machine.

4.1 Collect data

As we already know machine learning is data oriented. We need data to teach our system for future predictions.

4.2 Prepare the input data

Now you've downloaded the data, but when you're feeding the data, you need to make sure of the particular order of the data to make it meaningful for you machine learning tool to process it; i.e. CSV (comma separated value)file. This is the best format of the file to process the data. Because comma separated values help a lot in clustering.

4.3 Analyzing data

Now, you're looking at the patterns in the data to process it in a better way. You're checking the outliers (scope & boundaries) of the data. And you are also checking the novelty (specification) of the data.

4.4 Train model

This is the main part of the machine learning when you are developing the algorithm where you are structuring the complete system with coding to process the input and give back the output.

4.5 Test the model

Here you're checking the values you're getting from the system whether it matches your required outcome or not.

4.6 Deploy it in the application

Let's discuss an example of autonomous cars which don't have human intervention, which run on their own. The first step is to collect the data, and you have to collect many kinds of data. You're driving the car which runs on its owe. The car should know the road signs, it should have the knowledge of traffic signals and when people crossing the road, so that it can make the decision to stop or run in different situations. So we need a collection of images of these different situations, it is our collect data module.

Now we have to make the particular format of data (images) like csv file where we store the path of the file, the dimensions of the file. It makes our system processing efficient. This is what we called preparing the data.

And then it makes the patterns for different traffic signals (red, green, blue), for different sign boards of traffic and for its environment in which car or people running around. And then it decides the outlier of these objects whether it is static (stopped) or dynamic (running state). And then it can make the decision to stop or to move in the side of another object.

These decisions are obviously dependent upon training the model, and what code we write to develop our model. This is what we say training the model and then we test it and then we deploy it in our real world applications.

5. Applications of machine learning

Machine learning is widely used today in our applications.

- You might use the Snapchat or Instagram APP where you can apply the different animal's body parts on your face like ears, nose, tongue etc. These different organs places at the exact right spot in the image, this is an application of machine learning.
- Google is a widely used AI, ML. Google Lens is an application. If you scan anything through Google Lens then it can tell the properties and features of this specific thing.
- Google Maps is also using machine learning. For example, if you're watching any department

store on the map, sometimes it is telling you how much something is priced, and how expensive it is.

New Words

action	['ækʃn]	*n.*行动，活动；功能，作用
train	[treɪn]	*v.*训练；教育；培养
situation	[ˌsɪtju'eɪʃn]	*n.*（人的）情况；局面，形势，处境；位置
conclude	[kən'klu:d]	*vt.*得出结论；推断出；决定
comprehensive	[ˌkɒmprɪ'hensɪv]	*adj.*广泛的；综合的
bulk	[bʌlk]	*n.*大块，大量；大多数，大部分
scenario	[sə'nɑ:rɪəʊ]	*n.*设想；可能发生的情况；剧情梗概
backend	['bækend]	*n.*后端
random	['rændəm]	*adj.*任意的；随机的
		*n.*随意；偶然的行动
attract	[ə'trækt]	*vt.*吸引；引起……的好感（或兴趣）
		*vi.*具有吸引力；引人注意
function	['fʌŋkʃn]	*n.*函数
accurate	['ækjərət]	*adj.*精确的，准确的；正确无误的
validate	['vælɪdeɪt]	*vt.*确认；证实
knowledge	['nɒlɪdʒ]	*n.*知识；了解，理解
delete	[dɪ'li:t]	*v.*删除
update	[ˌʌp'deɪt]	*vt.*更新
	['ʌpdeɪt]	*n.*更新
gradually	['grædʒuəlɪ]	*adv.*逐步地，渐渐地
horizontally	[ˌhɒrɪ'zɒntəlɪ]	*adv.*水平地，横地
vertical	['vɜ:tɪkl]	*adj.*垂直的，竖立的
		*n.*垂直线，垂直面
approximately	[ə'prɒksɪmətlɪ]	*adv.*近似地，大约
reward	[rɪ'wɔ:d]	*n.*奖赏；报酬；赏金；酬金
		*vt.*奖赏；酬谢
optimal	['ɒptɪməl]	*adj.*最佳的，最优的；最理想的
feed	[fi:d]	*vt.*馈送；向……提供
comma	['kɒmə]	*n.*逗号
separated	['sepəreɪtɪd]	*adj.*分隔的，分开的
novelty	['nɒvltɪ]	*n.*新奇，新奇的事物
deploy	[dɪ'plɔɪ]	*v.*使展开；施展；有效地利用
intervention	[ˌɪntə'venʃn]	*n.*介入，干涉，干预
path	[pɑ:θ]	*n.*路径
decide	[dɪ'saɪd]	*vt.*决定；解决；裁决

		*vi.*决定；下决心
static	['stætɪk]	*adj.*静止的；不变的
organ	['ɔ:gən]	*n.*器官；元件

Phrases

as the name suggests	顾名思义
take a look into	看一看
data oriented	数据导向
be concerned with	涉及；与……有关
wireless communication	无线通信
knowledge set	知识集
straight line	直线
road sign	交通标志，路标
traffic signal	交通信号；红绿灯

Abbreviations

TB (TeraByte)	太字节

Word Building

1. 常见的构成动词的后缀

(1) -ate 表示：使……

elimination 排除，消除，消灭——eliminate 排除，消除，消灭

circulation 循环，流通——circulate（使）循环，（使）流通

termination 终止——terminate（使）终止

(2) -en 表示：使具有某种特性

wide 宽的——widen（使）变宽

dark 黑暗的——darken（使）变暗

strength 力量——strengthen 加强，巩固

weak 柔弱的，虚弱的——weaken 削弱，变弱

(3) -fy 表示：（使）成为；（使）……化

simple 简单——simplify 简化

specific 特定的——specify 具体说明

test 实验，测试——testify 证明，表明

beautiful 美丽的——beautify 美化

(4) -ize 表示：变成；（使某物更具有某种特点）

character 特征——characterize 以……为特征

emphasis 重要性，——emphasize 强调

real 真的—— realize 实现

industrial 工业的——industrialize 使工业化

2. 常见的构成副词的后缀

(1) -ly 加在名词后，表示每……地，每……时间一次地；加在形容词后，表示状态、程度、性质、方式。

day 日——daily 每日地

year 年——yearly 每年的，每年，一年一次

great 大的——greatly 大大地

quick 迅速地——quickly 迅速地

careful 小心的，仔细的——carefully 小心地，认真地

true 真正的，确实的——truly 真正地，确实地

(2) -wards 表示：向……

out 外边——outwards 向外

east 东——eastwards 向东

down 下——downwards 向下

upwards 向上——upwards 向上

(3) -wise 表示：方向；方式；状态

clock 钟，时钟——clockwise 顺时针方向

like 像——likewise 同样的

other 其他的——otherwise 否则

length 长，长度——lengthwise 纵长地

3. 常见的构成数词的后缀

(1) -th 表示：第……

five——fifth 第五

nine 九——ninth 第九

(2) -teen 表示：十……

five 五——fifteen 十五

nine 九——nineteen 十九

(3) -ty 表示：……十

eight 八 ——eighty 八十

six 六——sixty 六十

Career Training

论 文 摘 要

1. 摘要的概念和作用

摘要提供文献内容梗概，它简明、确切地记述文献的重要内容。其基本要素包括研究工作的主要对象和范围、采用的手段和方法、得出的结果和重要的结论，有时也包括具有情报价值的其他重要信息。摘要应具有独立性和自明性，并且拥有与文献相同的主要信息。即不阅读全文，就能获得必要的信息。

2. 英文摘要注意事项

英语有其独特的表达方式、语言习惯，在撰写英文摘要时应特别注意。

2.1　英文题名

（1）题名的结构　英文题名以短语为主要形式，尤以名词短语（noun phrase）最常见。即题名基本上由一个或几个名词加上其前置和（或）后置定语构成。例如，The Construction of Streaming Media-based English Learning Websites（基于流媒体的英语学习网站建设）。

短语型题名要先确定中心词，再进行前后修饰。词的顺序很重要，词序不当，会导致表达不准。题名一般不应是陈述句，因为题名主要起标示作用，而陈述句容易使题名具有判断性语义。

（2）题名的字数　题名不应过长。国外科技期刊一般对题名字数有限制。在能准确反映论文特定内容的前提下，题名字数越少越好。

（3）题名中的大小写　题名字母的大小写有以下 3 种格式：

- 全部字母大写。
- 每个词的首字母大写，但冠词、连词、介词全部小写。例如，Issues on Compiling Computer English Textbooks（试谈计算机英语教材编写的若干问题）。
- 题名第一个词的首字母大写，其余字母均小写。例如，The application of streaming media in English learning websites（流媒体在英语学习网站中的应用）。

（4）题名中慎用缩略词语。已得到整个科技界或本行业科技人员公认的缩略词语才可用于题名中，否则不要轻易使用。

2.2　英文摘要的时态

英文摘要时态的运用也以简练为佳。通常使用一般现在时、一般过去时，较少用现在完成时、过去完成时，进行时态和其他复合时态则基本不用。

一般现在时用于说明研究目的、叙述研究内容、描述结果、得出结论、提出建议或讨论等。涉及公认事实、自然规律、永恒真理等，也要用一般现在时。

一般过去时用于叙述过去某一时刻（时段）的发现、某一研究过程（实验、观察、调查等过程）。用一般过去时描述的发现、现象，往往是尚不能确认为自然规律、永恒真理的，而只是当时的情况，所描述的研究过程也明显带有过去时间的痕迹。

现在完成时把过去发生的或过去已完成的事情与现在联系起来。

过去完成时可用来表示过去某一时间以前已经完成的事情，或在一个过去事情完成之前就已完成的另一过去行为。

2.3　英文摘要的语态

采用何种语态，既要考虑摘要的特点，又要满足表达的需要。一篇摘要很短，尽量不要随便混用，更不要在一个句子里混用。

（1）主动语态　现在主张摘要中的谓语动词尽量采用主动语态的越来越多，因其有助于清晰、简洁及有力的文字表达。The author explores the present situation and the existing problems on the construction of computer English textbooks 比 The present situation and the existing problems on the construction of computer English textbooks are explored 语感要强。

（2）被动语态　以前强调多用被动语态，理由是科技论文主要是说明事实经过，至于这件事是谁做的，无须一一证明。事实上，在指示性摘要中，为强调动作承受者，还是采用被动语态为好。即使在报道性摘要中，有些情况下被动者无关紧要，也必须用强调的事物做主语。例如，In this case, a greater accuracy in measuring distance might be obtained。

2.4　英文摘要的人称

原来摘要的首句多用第三人称（This paper）等开头，现在倾向于采用更简洁的被动语态或原形动词开头。例如，To describe，To study，To investigate，To assess，To determine。

行文时最好不用第一人称，以方便文摘刊物的编辑刊用。

3. 避免常见错误

1）冠词错误　定冠词 the 易被漏用。the 用于表示整个群体、分类、时间、地名以外的独一无二的事物、形容词最高级等较少漏用，用于特指时常被漏用。这里有个原则，即在用 the 时，听者或读者已经确知所指的是什么。

2）数词错误　避免用阿拉伯数字做首词，如：Five thousand samples are collected…中的 Five thousand 不要写成 5000。

3）单复数错误　一些名词单复数形式不易辨认，从而造成谓语形式出错。

4）尽量使用短句　因为长句容易造成语义不清，所以应尽量使用短句。而且应尽量避免单调和重复。

4. 常用句式

虽然论文摘要的写法并非一成不变的，但初学者仍可以模仿一些常见的句子，下列句子都有参考价值。

4.1　开门见山，说明文章内容

The aim(or object, purpose) of this paper(or note) is to prove(show/present /develop/generalize /investigate)…

It is the purpose of this paper to prove(show/present/develop/generalize/investigate)…

This paper is concerned with…

This paper deals with…

In this paper we prove…

In this paper we present…

In this paper we propose to show…

4.2　回顾历史，阐述新见解

The problem…was first treated by…and later…improved by…

The purpose of this paper is to prove that it holds in a more general case.

… first raised the problem which was later partly solved by …

We now solve this problem in the case of …

4.3　对他人成果的改进

The purpose of this paper is to generalize the results obtained by …to a more general case, i. e. , …

In this paper we shall prove several theorems which are generalizations to the results given by …

This paper intends to remove some unnecessary assumptions (e. g. , regularity) from the paper on…

This paper deals with generalizations of the following problem…
This paper improves the result of …on…by weakening the conditions…

软件水平考试试题解析

【真题再现】

从供选择的答案中选出应填入下列英语文句中____内的正确答案，把编号写在答卷的对应栏内。

An antivirus program __A__ a virus by searching code recognized as that of one of the thousands of viruses known to afflict computer systems. An antivirus program also can be used to create a checksum for __B__ files on your disk, save the checksums in a special file, and then use the statistic to __C__ whether the files are modified by some new virus. Terminate and stay resident (TSR) programs can check for unusual __D__ to access vital disk areas and system files, and check files you copy into memory to be sure they are not __E__.

供选择的答案

A：① declares ②deducts ③ defeats ④ detects
B：① valuable ② variable ③ voluble ④ vulnerable
C：① calculate ②determine ③ run ④ write
D：① attempts ② objects ③ deprograms ④ routines
E：① copied ②effected ③ infected ④ injected

【答案】A：④ B：④ C：② D：① E：③

【试题解析】

A：declare 的意思是“宣布，宣告，声明”，deduct 的意思是“减去，扣除”，defeat 的意思是“打败，战胜”，detect 的意思是“检测，发现，探测”。防病毒程序的主要任务是检测病毒，发现系统或文件中的病毒，故选④。

B：本句意思是防病毒程序也可以对磁盘上一些脆弱文件建立检查统计。4 个选项中只有 vulnerable 符合题意，意思是“易受攻击的，脆弱的”，故选④。

C：calculate 的意思是“计算”，determine 的意思是“断定，判断，确定，测定”，run 的意思是“运行”，write 的意思是“写”。根据句意，选②。

D：attempt 的意思是“尝试，企图”，常与 to 连用，object 作名词的意思是“目标，对象，物体”。如：an object of admiration 的意思是赞美的对象，with no object in life 的意思是没有生活目标。如果做动词，它的意思是“反对”，常与 to 连用。如：I object to the proposal，意为我反对这个提议。deprogram 是一个动词，意思是，反洗脑“消除受毒化的思想”，routine 的意思是“例行程序”。根据句意，选①。

E：copy 的意思是“复制”，effect 的意思是“影响”，infect 的意思是“传染；感染”，inject 的意思是“注射”。根据句意，选③。

【参考译文】

防病毒程序通过查找已被发现的数千种能攻击计算机系统的病毒代码来检测病毒，同时对磁盘上一些脆弱文件建立检查统计，并把统计结果存入某个特殊文件中，然后利用这些统计数据判断文件是否已被某种新病毒修改过。特定的终端驻留程序（TSR）能检查一些不常见的对重要磁盘区和文件系统的访问，并且能够检查复制到内存的文件以确认它们没有被病毒感染。

Exercises

[Ex. 1] 根据 Text A 回答以下问题。

1) What is Artificial Intelligence? What do these processes include?

2) What do particular applications of AI include?

3) What can AI be categorized as? What are they respectively?

4) What does AI as a service allow individuals and companies to do?

5) What do researchers and marketers hope the label augmented intelligence will do?

6) How many types does Arend Hintze categorize AI into? What are they?

7) AI is incorporated into a variety of different types of technology. What are some examples mentioned in the passage?

8) How many types of machine learning algorithms are there? What are they?

9) What are the areas AI has made its way into?

10) What did the National Science and Technology Council (NSTC) do in 2016?

[Ex. 2] 根据 Text B 填空。

1) Machine learning means ____________________. It is ______________ through which we teach the machines about things. It is a branch of ________________.

2) We teach the machine with ____________________, ____________ and ______________. And then, we predict with the machine from these learnings and ______________.

3) Machine learning is closely related to ______________ and ______________. Machine learning is ____________.

4) In supervised learning, we know about __________________________ and we know ____________________.

5) Because we've started with known values for our inputs, we can ____________ and ____________.

6) In unsupervised learning, we really don't know ____________.

7) Reinforcement learning is like ____________. It is also ____________.

8) There are some key point steps of machine learning when we start to teach the machine. They are ________, ________, ________, ________, ________, and ________.

9) ____________ is the best format of the file to process the data. Because comma separated values help a lot in ____________.

10) Train model is the main part of the machine learning when you are ____________ where you are structuring the complete system with coding to ____________ and ____________.

[Ex. 3] 把下列句子翻译为中文。

1) The only rule that matters is do what feels right to you.

2) This reasoning should be capped by an experimental finding.

3) But today, computer vision is not in that state yet.

4) The digital signature scheme can solve this issue .

5) You should always run virus detection software.

6) There are abnormal running present sometimes in computer network, we have to call some commands of network testing to detect and diagnose the failure.

7) The hypothesis must check with the facts.

8) To optimize performance of an application, you should attempt to minimize wait time at each component.

9) It is difficult to predict how long it will take to develop human-level artificial general intelligence.

10) A speech recognition system must expend significant processing to turn your speech into text.

[Ex. 4] 软件水平考试真题自测。

从供选择的答案中选出应填入下列英语文句中____内的正确答案，把编号写在答卷的对应栏内。

Communication protocols are __A__ connection-oriented or connectionless, __B__ whether the sender of a message needs to contact and maintain a dialog with the recipient or __C__ send a

message without any prior connect and with the hope that the recipient receives everything ____D____.
These methods ____E____ the two ways that communication is implemented on networks.

The major problem with e-mail is that it is ____F____ easy to use that people can become ____G____ with messages ____H____ they can possibly answer in a day. In addition, mail boxes require some management to ____I____ messages or archive those that might be required later. Senders don't always know about your E-mail backlog and often send ____J____ messages.

供选择的答案：

A：①not ②neither ③either ④all

B：①filled ②flooded ③depending on ④defined by

C：①immediately ②simply ③accordingly ④properly

D：①in order ②in array ③in series ④in queuing

E：①make known ②disclose ③reveal ④discover

F：①too ②so ③very ④much

G：①full ②lost ③inundated ④filled

H：①more than ②than ③that ④which

I：①mange ②save ③backup ④dispose of

J：①too many ②redundant ③long ④trivial

[Ex. 5] 听短文填空。

Exercises 12

An expert system is a computer program that uses artificial intelligence (AI) technologies to ____1____ the judgment and behavior of a human or an organization that has ____2____ knowledge and experience in a particular ____3____.

Typically, an expert system incorporates knowledge base containing accumulated ____4____ and an inference or rules engine—a set of rules for applying the knowledge base to each particular situation that is described to the program. The system's capabilities can be ____5____ with additions to the knowledge base or to the set of rules. Current systems may include machine ____6____ capabilities that allow them to improve their performance based on experience, just as humans do.

Expert systems have played a large role in many ____7____ including in financial services, telecommunications, health care, customer service, ____8____, video games, manufacturing, aviation and written communication. Two early expert systems broke ground in the health care space for ____9____ diagnoses: Dendral, which helped chemists identify ____10____ molecules, and MYCIN, which helped to identify bacteria such as bacteremia and meningitis, and to recommend antibiotics and dosages.

Reference Translation

人 工 智 能

人工智能（AI）是机器特别是计算机系统对人类智能处理的模拟。这些过程包括学习

（获取信息和使用信息的规则）、推理（使用规则来得出近似或明确的结论）和自我校正。人工智能的典型应用包括专家系统、语音识别和机器视觉。

人工智能可以分为弱人工智能与强人工智能两类。弱人工智能，也被称为窄人工智能，是为特定任务而设计和训练的人工智能系统。虚拟个人助理，如 Apple 的 Siri，就是一种弱人工智能。强人工智能，也被称为通用人工智能，是一种具有广泛人类认知能力的人工智能系统。当提出一项不熟悉的任务时，强人工智能系统能够在没有人为干预的情况下找到解决方案。

由于人工智能的硬件、软件和人员成本可能很昂贵，因而许多供应商在其标准产品中包含人工智能组件以及对人工智能即服务（AIaaS）平台的访问功能。在做出投入之前，人工智能即服务允许个人和公司为各种商业目的进行人工智能试验，并对多个平台进行抽样调查。现在流行的人工智能云产品包括 Amazon AI 服务、IBM Watson Assistant、Microsoft Cognitive Services 和 Google AI 服务。

虽然人工智能工具为企业提供了一系列新功能，但人工智能的使用也引发了伦理问题。这是因为深度学习算法作为许多最先进的人工智能工具的基础，其智能仅仅与训练时所提供的数据匹配。人类选用何种数据来训练人工智能程序，可能带有人类固有的偏见，因此必须密切监控。

一些业内专家认为，人工智能这一术语与流行文化联系太紧密，导致普通大众对人工智能产生不切实际的恐惧，以及对人工智能如何改变工作场所和生活方式抱有不太可能的期望。研究人员和营销人员希望增强智能（具有更中性内涵）可以帮助人们明白人工智能只会简单地改进产品和服务，而不是取代使用它们的人。

1. 人工智能的类型

密歇根州立大学综合生物学和计算机科学与工程助理教授 Arend Hintze 将人工智能分为 4 类，从现有的人工智能系统到尚未存在的感觉系统。他的分类如下：

类型 1：反应机器。例如，Deep Blue（深蓝），一个在 20 世纪 90 年代击败 Garry Kasparov 的 IBM 国际象棋程序。Deep Blue 可以识别棋盘上的棋子并进行预测，但它没有记忆，也无法使用过去的经验来指导未来的棋子。它分析了可能的行动——它自己和它的对手——并选择最具战略性的举措。Deep Blue 和 Google 的 AlphaGO 专为特定目的而设计，不能轻易应用于其他情况。

类型 2：有限的存储。这些人工智能系统可以使用过去的经验来指导未来的决策。自动驾驶汽车的一些决策功能就是这样设计的。观察可以为在不远的将来发生的行动提供信息，例如，换车道。这些观察结果不会永久存储。

类型 3：心智理论。这个心理学术语指的是理解他人有自己的信念、欲望和意图，这会影响他们的决策。这类人工智能尚不存在。

类型 4：自我意识。在这个类型中，人工智能系统具有自我意识，具有知觉。具有自我意识的机器了解其当前状态，并可以使用状态信息来推断其他人的感受。这种类型的人工智能尚不存在。

2. 人工智能技术的例子

人工智能被整合到各种不同类型的技术中。这里有一些例子。

自动化：可以使系统或过程自动运行。例如，机器人过程自动化（RPA）可以通过编程来执行人类通常执行的大量、可重复的任务。RPA 与 IT 自动化的不同之处在于它可以适应不断变化的环境。

机器学习：使计算机无须编程即可行动的技术。深度学习是机器学习的一个子集，简言之，它可以被认为是通过自动化进行预测分析。有三种类型的机器学习算法：

- 监督学习：标记数据集，以便可以检测模式并用于标记新数据集。
- 无监督学习：不标记数据集，并根据相似性或差异性进行分类。
- 强化学习：不标记数据集，但在执行一个行动或多个行动后，人工智能系统会得到反馈。

机器视觉：让计算机具有视觉的技术。该技术使用相机、模数转换和数字信号处理来捕获和分析视觉信息。它通常被比作人类的视觉，但机器视觉不受生物学的约束，并且可以编程以透视墙壁。它用于从签名识别到医学图像分析的各种应用中。计算机视觉聚焦基于计算机的图像处理，通常与机器视觉混为一谈。

自然语言处理（NLP）：通过计算机程序处理人类——而不是计算机——的语言。其中一个较早且最著名的 NLP 示例是垃圾邮件检测，它查看主题和电子邮件的文本并确定此邮件是否为垃圾邮件。目前的 NLP 方法是基于机器学习的。NLP 任务包括文本翻译、情感分析和语音识别。

机器人技术：一个专注于机器人设计和制造的工程技术。机器人通常用于执行人类难以执行或一直执行的任务。它们用于汽车生产的装配线或被 NASA 用于在太空中移动大型物体。研究人员还利用机器学习来构建可在社交场合进行交互的机器人。

自动驾驶汽车：它们把计算机视觉、图像识别和深度学习相结合，从而建立在指定车道上驾驶车辆并避开意外障碍物（如行人）的自动驾驶技能。

3. 人工智能应用

人工智能已经进入了许多领域。这里列举 6 个应用示例。

人工智能应用于医疗保健领域。最大的好处是改善对患者的治疗效果和降低成本。公司正在应用机器学习来做出比人类更好、更快的诊断。IBM Watson 是最著名的医疗保健技术之一。它理解自然语言，并能够回答所提出的问题。它挖掘患者数据和其他可用数据源以形成假设，然后给出一个置信评分模式。医疗保健领域的其他人工智能应用程序包括聊天机器人。聊天机器人是一个计算机程序，是用于在线回答问题和帮助客户、帮助安排后续预约或自动计费，以及提供基本医疗反馈的虚拟健康助理。

人工智能应用于商业领域。机器人流程自动化正被应用于通常由人类执行的高度重复的任务。机器学习算法正被集成到分析学和客户关系管理平台，以发现信息并更好地为客户服务。聊天机器人已被网站使用，为客户提供即时服务。工作岗位的自动化也成为学术界和 IT 分析师的话题。

人工智能应用于教育领域。人工智能可以自动评分，节省教师时间。人工智能可以评估学生并应对他们的需求，帮助他们按照自己的进度学习。人工智能导师可以为学生提供额外的支持，确保他们处于正确轨道上。人工智能可以改变学生学习的地点和方式，甚至可以取代一些教师。

人工智能应用于金融领域。个人理财应用程序中的人工智能（如 Mint 或 Turbo Tax）正在进入金融机构。这些应用程序收集个人数据并提供财务建议。其他程序（例如 IBM Watson）已经应用于购买房屋。今天，华尔街很大一部分交易都是由软件完成的。

人工智能应用于法律领域。在法律上，人工发现过程（筛选文件）是非常困难的工作。自动化地完成此项工作可以大大节省时间。创业公司还在构建计算机回答助手，通过检查与数据库相关的分类和本体，可以筛选出问题的答案。

人工智能应用于制造业。这个领域一直处于将机器人纳入工作流程的最前沿。过去工业机器人执行单一任务并与人类工作人员分开，但随着技术的进步这一现象已经发生了变化。

4．安全和伦理问题

人工智能在自动驾驶汽车领域的应用引发了安全和伦理方面的担忧。汽车可能被黑客入侵，当自动驾驶汽车发生事故时，责任是不清楚的。自动驾驶汽车也可能处于无法避免事故的情况，迫使编程人员就如何最大限度地减少损失做出伦理决定。

另一个主要担忧是滥用人工智能工具的可能性。黑客们开始使用复杂的机器学习工具来访问敏感系统，使安全问题越来越复杂化。

基于深度学习的视频和音频生成工具也为不良行为者提供了必要的工具来创建所谓“换脸术”。他们可以编造公众人物的视频，尽管这些公众人物从未说过这些话或做过这些事，但这些视频却让人们信以为真。

5．人工智能技术的规范

尽管存在这些潜在的风险，但很少有关于使用人工智能工具的法规，而且即便有一些法规，它们通常也只是间接地涉及人工智能。例如，联邦公平贷款条例要求金融机构向潜在客户解释信用决策，这些条例限制了贷方使用深度学习算法的程度，而这些算法本质上通常是不透明的。欧洲的《通用数据保护条例》（GDPR）严格限制了企业使用消费者数据的方法，这阻碍了许多面向消费者的人工智能应用程序的训练和功能。

2016 年，美国国家科学技术委员会（NSTC）发布了一份报告，研究政府监管在人工智能发展中可能发挥的作用，但并未建议考虑具体立法。从那时起，这个议题就很少受到立法者的关注。

附录

附录A 单 词 表

A

ability[ə'bɪlətɪ]*n.*能力，资格；能耐，才能

absorb[əb'sɔ:b]*vt.*吸收

abstract['æbstrækt]*adj.*抽象的，理论的*n.*摘要，概要，抽象

abstraction[æb'strækʃn]*n.*抽象，提取

accelerate[ək'seləreɪt]*v.*加速，促进

accelerometer[əkˌselə'rɒmɪtə]*n.*加速度计

access['ækses]*n.*访问，入门；*vt.*存取

accessibility[əkˌsesə'bɪlətɪ]*n.*可访问性

accessible[ək'sesəbl]*adj.*可访问的

accessory[ək'sesərɪ]*n.*附件，零件，附加物
*adj.*附属的，补充的，副的

accident['æksɪdənt]*n.*意外事件；事故

accordingly[ə'kɔ:dɪŋlɪ]*adv.*相应地；照着，依据

account[ə'kaʊnt]*n.*账号

accurate['ækjərət]*adj.*精确的，准确的；正确无误的

achievable[ə'tʃi:vəbl]*adj.*做得成的，可完成的

achieve[ə'tʃi:v]*vt.*取得，获得；实现，成功
*vi.*达到预期的目的，实现预期的结果

achievement[ə'tʃi:vmənt]*n.*成就，功绩

acknowledgement[ək'nɒlɪdʒmənt]*n.*确认，承认

acquisition[ˌækwɪ'zɪʃn]*n.*获得

action['ækʃn]*n.*行动，活动；功能，作用

active['æktɪv]*adj.*主动的，活动的

activity[æk'tɪvətɪ]*n.*行动，行为

actual['æktʃuəl]*adj.*实际的，真实的，现行的

adapter[ə'dæptə]*n.*适配器，转接器

address[ə'dres]*n.*地址

administer[əd'mɪnɪstə]*v.*管理，给予，执行

administrator[əd'mɪnɪstreɪtə]*n.*管理员

advantage[əd'vɑ:ntɪdʒ]*n.*优势，优点，有利条件

advocate['ædvəkeɪt]*n.*提倡者，支持者；*vt.*提倡，支持

aerospace['eərəuspeɪs]*n.*航空航天

affiliate[ə'fɪlieɪt]*v.*(使……)加入，被接受为会员

affordable[ə'fɔ:dəbl]*adj.*买得起的；价格合理的；负担得起的

agency['eɪdʒənsɪ]*n.*代理；机构

agile['ædʒaɪl]*adj.*敏捷的，轻快的，灵活的

aid[eɪd]*n.&vt.*辅助，帮助

algebra['ældʒɪbrə]*n.*代数学

algorithm['ælgərɪðəm]*n.*算法；运算法则

alphanumeric[ˌælfənju:'merɪk]*adj.*字符数字的，字母数字混合编制的

alternative[ɔ:l'tɜ:nətɪv]*n.*可供选择的办法（事物）
*adj.*选择性的

amaze[ə'meɪz]*vt.*使大为吃惊，使惊奇
*n.*吃惊，好奇

ambient['æmbɪənt]*adj.*周围的，环境的

ambiguous[æm'bɪgjuəs]*adj.*不明确的

amend[ə'mend]*v.*改良，修改；修订

amplify['æmplɪfaɪ]*vt.*放大，增强

analogous[ə'næləgəs]*adj.*相似的，可比拟的

analyst['ænəlɪst]*n.*分析师

analytical[ˌænə'lɪtɪkl]*adj.*分析的，解析的

analytics[ˌænə'lɪtɪks]*n.*分析学，解析学，分析论

analyze['ænəlaɪz]*vt.*分析，分解

Android['ændrɔɪd]*n.*安卓

anecdote['ænɪkdəʊt]*n.*趣闻；逸事

annoying[ə'nɔɪɪŋ]*adj.*恼人的，讨厌的

anomalous[ə'nɒmələs]*adj.*不规则的，反常的

anomaly[ə'nɒməlɪ]*n.*不正常的

anonymous[ə'nɒnɪməs]adj.匿名的
apparel[ə'pærəl]n.衣服，装饰
applet['æplət]n.小应用程序
application[ˌæplɪ'keɪʃn]n.应用，应用程序，应用软件
appoint[ə'pɔɪnt]vt.任命；委派
appointment[ə'pɔɪntmənt]n.约定
approach[ə'prəʊtʃ]n.方法，步骤，途径
appropriate[ə'prəʊprɪət]adj.适当的
approve[ə'pru:v]v.批准，认可，通过
approximate[ə'prɒksɪmət]adj.极相似的；vi.接近于，近似于；vt.靠近，使接近
approximately[ə'prɒksɪmətlɪ]adv.近似地，大约
archaeologist[ˌɑ:kɪ'ɒlədʒɪst]n.考古学家
architecture['ɑ:kɪtektʃə]n.体系结构
arithmetic[ə'rɪθmətɪk]n.算术，算法
arrange[ə'reɪndʒ]v.安排，排列
arrangement[ə'reɪndʒmənt]n.排列，安排
array[ə'reɪ]n.数组
artifact['ɑ:tɪfækt]n.人工制品，手工艺品，加工品
artificial[ˌɑ:tɪ'fɪʃl]adj.人造的
assembler[ə'semblə]n.汇编程序
assign[ə'saɪn]vt.分配，分派，选派
assignment[ə'saɪnmənt]n.赋值，分配
association[əˌsəʊʃɪ'eɪʃn]n.协作，联合
asynchronous[eɪ'sɪŋkrənəs]adj.异步的
attacker[ə'tækə]n.攻击者
attention[ə'tenʃn]n.注意，注意力
attenuation[əˌtenju'eɪʃn]n.衰减
attract[ə'trækt]vt.吸引；引起……的好感（或兴趣）vi.具有吸引力；引人注意
attribute[æ'trɪbju:t]n.属性，品质，特征
audio['ɔ:dɪəʊ]adj.音频的，声频的，声音的
authenticate[ɔ:'θentɪkeɪt]v.鉴别
author['ɔ:θə]vt.写作，创作；n.作家，创造者
authority[ɔ:'θɒrətɪ]n.权威，威信
authorization[ˌɔ:θəraɪ'zeɪʃn]n.授权，认可
automate['ɔ:təmeɪt]vt.自动化，使自动操作
autonomy[ɔ:'tɒnəmi]n.自治
available[ə'veɪləbl]adj.可用到的，可利用的，有用的
available[ə'veɪləbl]adj.可利用的；可获得的；能找到的

B

backbone['bækbəʊn]n.中枢，骨干
backend['bækend]n.后端
ballistics[bə'lɪstɪks]n.弹道学，发射学
bankrupt['bæŋkrʌpt]adj.破产的，倒闭的；n.破产者；vt.使破产
bankruptcy['bæŋkrʌptsi]n.破产
batch[bætʃ]n.批处理
behavior[bɪ'heɪvjə]n.举止，行为
believable[bɪ'li:vəbl]adj.可信的
Berlin[bə:'lin]n.柏林(德国城市)
bill[bɪl]n.钞票，账单，票据，清单
binary['baɪnərɪ]adj.二进制的，二进位的
bind[baɪnd]vt.绑定
bit[bɪt]n.位，比特（二进制数字4的位，信息量的度量单位）
blog[blɒg]n.博客
bombard[bɒm'bɑ:d]vt.炮轰，轰击
bookmark['bʊkmɑ:k]n.书签；vt.给……设置书签
boolean['bu:lɪən]adj.布尔的
boost[bu:st]v.推进
boot[bu:t]v.引导，导入
boundary['baʊndrɪ]n.边界，分界线，界限，范围
branch[brɑ:ntʃ]n.分支
breach[bri:tʃ]n.破坏，裂口；vt.打破，突破
bribe[braɪb]v.贿赂，行贿；n.贿赂
broadband['brɔ:dbænd]n.宽带
broadcast['brɔ:dkɑ:st]n.广播，播音；v.广播，播送，播放
broker['brəʊkə]n.经纪人
brouter['bru:tə]n.网桥路由器
browser['braʊzə]n.浏览器，浏览程序
Budapest['bu:dəpest]n.布达佩斯（匈牙利首都）
budget['bʌdʒɪt]n.预算；vi.做预算，编入预算
buffer['bʌfə]n.缓冲器
bug[bʌg]n.程序缺陷
built-in[bɪlt-ɪn]adj.内置的，固定的，嵌入的n.内置
bulk[bʌlk]n.大块，大量；大多数，大部分

bundle['bʌndl]n.捆，束，包；v.捆扎
bury['beri]vt.掩埋，隐藏
bus[bʌs]n.总线
buzzword['bʌzwɜ:d]n.（报刊等的）时髦术语，流行词
byte[baɪt]n.字节
bytecode['baɪtkəʊd]n.字节码

C

cable['keɪbl]n.电缆，缆，索；电缆电报
capability[ˌkeɪpə'bɪlətɪ]n.能力；才能；才干
capacity[kə'pæsɪtɪ]n.容量
capture['kæptʃə]vt.捕获
card[kɑ:d]n.插卡
careful['keəfl]adj.小心的，仔细的
category['kætəgərɪ]n.种类，别
certification[ˌsɜ:tɪfɪ'keɪʃn]n.证明，认证
channel['tʃænl]n.通道，渠道
character['kærəktə]n.字符
characterize['kærəktəraɪz]vt.表现……的特色
chart[tʃɑ:t]n.图表；vt.制图
chatbot[tʃætbɒt]n.聊天机器人
chip[tʃɪp]n.芯片
churn[tʃɜ:n]v.流失
circuit['sɜ:kɪt]n.电路，一圈
circumstance['sɜ:kəmstəns]n.环境，境遇
claim[kleɪm]vt.（根据权利）要求，索赔
clarify['klærəfaɪ]v.澄清，阐明
clearance['klɪərəns]n.证明书无过失、可靠或称职的官方证明，批准证
clickstream['klɪkstri:m]n.点击流量
clockwise['klɒkwaɪz]adj.顺时针方向的
adv.顺时针方向地
cloudburst['klaʊdbɜ:st]n.云破裂
cloudware['klaʊdweə]n.云件
cluster['klʌstə]n.串，丛；vi.丛生，成群
clustering['klʌstərɪŋ]n.聚类
coaxial[kəʊ'æksɪəl]adj.同轴的，共轴的
cockpit['kɒkpɪt]n.驾驶员座舱
code[kəʊd]n.代码；vt.编码
coder['kəʊdə]n.写代码的人，编码者
coerce[kəʊ'ɜ:s]vt.控制，限制；威胁；逼迫
cognitive['kɒgnətɪv]adj.认知的，认识的
collaborative[kə'læbərətɪv]adj.合作的，协作的
collect[kə'lekt]v.收集，聚集，集中，搜集
collection[kə'lekʃn]n.收集，收取
collision[kə'lɪʒn]n.碰撞，冲突
column['kɒləm]n.列，栏
combination[ˌkɒmbɪ'neɪʃn]n.结合，联合，合并
comma['kɒmə]n.逗号
commitment[kə'mɪtmənt]n.承诺，许诺；委任，委托
communication[kəˌmju:nɪ'keɪʃn]n.通信
community[kə'mju:nətɪ]n.社区，团体
compact[kəm'pækt]adj.紧凑的；简洁的
competition[ˌkɒmpɪ'tɪʃn]n.竞争者；竞争，竞赛
competitive[kəm'petətɪv]adj.竞争的
compilation[ˌkɒmpɪ'leɪʃn]n.编译
compile[kəm'paɪl]vt.编译，编辑，汇编
compiler[kəm'paɪlə]n.编译器，编译程序
complete[kəm'pli:t]adj.完整的；完全的；完成的
vt.完成，使完满
complex['kɒmpleks]adj.复杂的，合成的，综合的
n.联合体
complication[ˌkɒmplɪ'keɪʃn]n.纠纷；混乱
component[kəm'pəʊnənt]n.成分，部件
adj.组成的，构成的
compose[kəm'pəʊz]v.组成
composite['kɒmpəzɪt]adj.合成的，复合的；n.合成物
comprehensive[ˌkɒmprɪ'hensɪv]adj.广泛的；综合的
compression[kəm'preʃn]n.压缩
comprise[kəm'praɪz]v.包含，由……组成
compromise['kɒmprəmaɪz]vt.违背（原则）；（尤指因行为不很明智）使陷入危险
computation[ˌkɒmpju'teɪʃn]n.计算，估计
computer[kəm'pju:tə]n.（电子）计算机，电脑
computerize[kəm'pju:təraɪz]vt.用计算机处理，使计算机化
con[kɒn]adv.反对地，反面地
n.反对的意见；反对票，缺点
concentrate['kɒnsntreɪt]v.专心于；注意；集中

concentrator['kɒnsentreɪtə]n.集中器
concept['kɒnsept]n.概念，观念
conceptual[kən'septʃuəl]adj.概念上的
conceptually[kən'septʃuəlɪ]adv.概念地
conclude[kən'klu:d]vt.得出结论；推断出；决定
conclusion[kən'klu:ʒn]n.结论；断定，决定；推论
concrete['kɒŋkri:t]adj.具体的，有形的
concurrency[kən'kʌrənsɪ]n.并发（性）
concurrently[kən'kʌrəntlɪ]adv.同时地
condition[kən'dɪʃn]n.条件，情形，环境
conditional[kən'dɪʃənl]adj.条件的，假定的
conduit['kɒndjuɪt]n.管道，导管
confidentiality[ˌkɒnfɪˌdenʃɪ'ælətɪ]n.机密性
configurable[kən'fɪgərəbl]adj.结构的，可配置的
configuration[kənˌfɪgə'reɪʃn]n.构造，结构，配置
congestion[kən'dʒestʃən]n.拥塞
connect[kə'nekt]v.连接，联合
connotation[ˌkɒnə'teɪʃn]n.内涵，含义
consciousness['kɒnʃəsnəs]n.意识，观念；知觉
consequence['kɒnsɪkwəns]n.结果，推论，因果关系
consistently[kən'sɪstəntlɪ]adv.一贯地，坚持地
console[kən'səʊl]n.控制台，操纵台
consolidate[kən'sɒlɪdeɪt]vt.把……合成一体，合并
vi.合并，联合
constitute['kɒnstɪtju:t]vt.构成，组成；制定，设立
constraint[kən'streɪnt]n.强制，约束
construct[kən'strʌkt]vt.建造，构造，创立
consumption[kən'sʌmpʃn]n.消费
contentious[kən'tenʃəs]adj.争论的，有异议的
continually[kən'tɪnjuəlɪ]adv.不停地；持续地；屡屡地；一再地
continuity[ˌkɒntɪ'nju:ətɪ]n.连续性，连贯性
contract['kɒntrækt]n.合同，契约
contradictory[ˌkɒntrə'dɪktərɪ]adj.反驳的，反对的
n.矛盾因素，对立物
control[kən'trəʊl]n.&vt.控制，支配，管理，操纵
controller[kən'trəʊlə]n.控制器
conversion[kən'vɜ:ʃn]n.变换，转化
convert[kən'vɜ:t]vt.使转变，转换……
convey[kən'veɪ]vt.传达，传递；运送，输送；表达
convincingly[kən'vɪnsɪŋlɪ]adv.令人信服地，有说服力地
cooperative[kəʊ'ɒpərətɪv]adj.合作的；协助的；共同的
correlate['kɒrəleɪt]v.使互相关联；联系；adj.相关的
correlation[ˌkɒrə'leɪʃn]n.相互关系，相关(性)
counteract[ˌkaʊntər'ækt]vt.抵消，阻碍
counterrotating['kaʊntərəʊ'teɪtɪŋ]adj.反向旋转的
counterweight['kaʊntəweɪt]n.平衡物，平衡锤，平衡力
cover['kʌvə]vt.包括，包含，适用
cracker['krækə]n.骇客，解密高手
craft[krɑ:ft]n.手艺；vt.手工制作；精巧地制作
credit['kredɪt]n.信誉，信用；[金融]贷款；vt.相信，信任
cripple['krɪpl]v.使瘫痪
criteria[kraɪ'tɪərɪə]n.标准
crucial['kru:ʃl]adj.至关紧要的
crude[kru:d]adj.粗糙的，简陋的
crunch[krʌntʃ]v.嘎扎嘎扎地咬嚼，压碎，扎扎地踏过
cryptography[krɪp'tɒgrəfɪ]n.密码学，密码术
crypto-shredding['krɪptəʊ-'ʃredɪŋ]n.密码粉碎
customization['kʌstəmaɪzeɪʃn]n.用户化，专用化，定制
customize['kʌstəmaɪz]v.定制，用户化
cutting-edge['kʌtɪŋedʒ]adj.前沿的

D

damage['dæmɪdʒ]n.损害，损毁；赔偿金
v.损害，毁坏
dangerous['deɪndʒərəs]adj.危险的
data['deɪtə]n.数据
database['deɪtəbeɪs]n.数据库
datacenter['deɪtəˌsentə]n.数据中心
deal[di:l]vi.处理，应付
debate[dɪ'beɪt]n.讨论；辩论；争论
debug[di:'bʌg]vt.调试程序，排除故障
decide[dɪ'saɪd]vt.决定；解决；裁决；vi.决定；下决心
decision[dɪ'sɪʒn]n.决策
declare[dɪ'kleə]vt.声明
decode[di:'kəʊd]vt.解码
decoy['di:kɔɪ]vt.诱骗
decryption[di:'krɪpʃn]n.解密，译码
dedicate['dedɪkeɪt]vt.提供

dedicated['dedɪkeɪtɪd]*adj.*专门的，专用的
deepfake['di:pfeɪk]*n.*"换脸术"，一种人物图像合成技术
default[dɪ'fɔ:lt]*n.*默认（值），缺省（值）
defect['di:fekt]*n.*过失，缺点
define[dɪ'faɪn]*vt.*定义，详细说明
definite['defɪnət]*adj.*明确的；一定的；肯定
degradation[ˌdegrə'deɪʃn]*n.*降级，降格，退化
degrade[dɪ'greɪd]*v.*退化，（使）降级，（使）退化
delete[dɪ'li:t]*v.*删除
deliberate[dɪ'lɪbərət]*adj.*深思熟虑的，故意的，预有准备的
deliver[dɪ'lɪvə]*vt.*发表；交付
delivery[dɪ'lɪvərɪ]*n.*递送，交付；发送，传输
demodulate[di:'mɒdjʊleɪt]*vt.*解调
demonstrate['demənstreɪt]*vt.*证明，论证
denial[dɪ'naɪəl]*n.*否认，否定，拒绝
deny[dɪ'naɪ]*v.*否认，拒绝
deploy[dɪ'plɔɪ]*v.*使展开；施展；有效地利用
description[dɪ'skrɪpʃn]*n.*描述，形容；种类，类型
desirability[dɪˌzaɪərə'bɪlətɪ]*n.*愿望，希求
desirable[dɪ'zaɪərəbl]*adj.*值得要的，合意的，称心如意的
destine['destɪn]*vt.*注定，预定
detail-oriented['di:teɪl-'ɔ:rɪəntɪd]*adj.*细节导向的，面向细节的，注重细节的
detect[dɪ'tekt]*vt.*察觉，发现，探测
detection[dɪ'tekʃn]*n.*检查，检测
deterministic[dɪˌtɜ:mɪ'nɪstɪk]*adj.*确定性的
developer[dɪ'veləpə]*n.*开发者
development[dɪ'veləpmənt]*n.*开发，发展
device[dɪ'vaɪs]*n.*装置，设备
diagnose['daɪəgnəʊz]*vt.*诊断；判断；*vi.*做出诊断
dial-up['daɪəlʌp]*v.*拨号（上网）
dictate[dɪk'teɪt]*vt.*命令，指示；控制，支配；
dig[dɪg]*v.*掘，挖，搜集
dimension[daɪ'menʃn]*n.*尺度，维(数)
diploma[dɪ'pləʊmə]*n.*文凭，毕业证书
directory[də'rektərɪ]*n.*目录
disallow[ˌdɪsə'laʊ]*vt.*不准许，禁止，不接受
disappear[ˌdɪsə'pɪə]*vi.*不见，不显示，消失；不复存在
discern[dɪ'sɜ:n]*v.*认识，洞悉，辨别，看清楚
discount['dɪskaʊnt]*n.*折扣
discourage[dɪs'kʌrɪdʒ]*vt.*阻碍
discover[dɪ'skʌvə]*vt.*发现，发觉
discrepancy[dɪs'krepənsi]*n.*矛盾；不符合（之处）
discriminate[dɪ'skrɪmɪneɪt]*v.*区别
disjoint[dɪs'dʒɔɪnt]*v.*（使）脱节，（使）解体
disk[dɪsk]*n.*磁盘
disparage[dɪ'spærɪdʒ]*vt.*贬低，毁谤，轻视
disparate['dɪspərət]*adj.*全异的
display[dɪ'spleɪ]*vi.*（计算机屏幕上）显示；*n.*显示，显示器
disposal[dɪ'spəʊzl]*n.*（事情的）处置；清理 *adj.*处理（或置放）废品的
dispose[dɪ'spəʊz]*v.*处理
dispute[dɪ'spju:t]*v.*& *n.*争论，辩论，争吵
dissertation[ˌdɪsə'teɪʃn]*n.* (学位)论文
distinct[dɪ'stɪŋkt]*adj.*清楚的，明显的，有区别的
distinction[dɪ'stɪŋkʃn]*n.*区别，差别
distinguish[dɪ'stɪŋgwɪʃ]*vi.*区分，辨别，分清
distract[dɪ'strækt]*v.*转移
distraction[dɪ'strækʃn]*n.*干扰，注意力分散
distribute[dɪ'strɪbju:t]*vt.*分发，分布
distributed[dɪs'trɪbju:tɪd]*adj.*分布式的
domestically[də'mestɪklɪ]*adv.*国内地；适合国内地
dominant['dɒmɪnənt]*adj.*占优势的；统治的，支配的
download[ˌdaʊn'ləʊd]*v.*下载
dramatically[drə'mætɪkəlɪ]*adv.*显著地，剧烈地
driver['draɪvə]*n.*驱动器，驱动程序
drug[drʌg]*n.*药，麻药；*vi.*吸毒；*vt.*使服毒品，毒化
duplicate['dju:plɪkeɪt]*adj.*复制的，两重的，两倍的，完全相同；*n.*复制品，副本；*vt.*复写，复制，使加倍，使成双
durable['djʊərəbl]*adj.*耐用的，耐久的；持久的；长期的 *n.*耐用品
dynamic[daɪ'næmɪk]*adj.*动态的；*n.*动态，动力
dynamical[daɪ'næmɪkəl]*adj.*动态的

dynamically[daɪ'næmɪklɪ]adv.动态地

E

earn[ɜ:n]v.赚得，获得

e-commerce[i:'kɒmɜ:s]n.电子商务

editor['edɪtə]n.编辑器

efficiency[ɪ'fɪʃnsɪ]n.效率，功效

elaborate[ɪ'læbəreɪt]vi.详尽说明；变得复杂

vt.详细制定；尽心竭力地做

elasticity[ˌi:læ'stɪsətɪ]n.弹性

electromechanical[ɪˌlektrəʊmɪ'kænɪkəl]adj.[机]电动机械的，机电的，电机的

elegant['elɪgənt]adj.简练的，简洁的；漂亮的

element['elɪmənt]n.要素，元素，成分，元件

elementary[ˌelɪ'mentrɪ]adj.初步的，基本的

eliminate[ɪ'lɪmɪneɪt]vt.排除，消除，除去

email['i:meɪl]n.电子邮件

embedded[ɪm'bedɪd]adj.嵌入的，植入的，内含的

emerge[ɪ'mɜ:dʒ]vi.显现，浮现，暴露，形成

emphasize['emfəsaɪz]vt.强调，着重

emulate['emjuleɪt]vt.仿真

emulation[ˌemju'leɪʃn]n.仿真，模拟

enable[ɪ'neɪbl]vt.使能够

encapsulate[ɪn'kæpsjuleɪt]v.包装，封装

encapsulation[inˌkæpsju'leiʃn]n.包装，封装

encode[ɪn'kəʊd]vt.编码

encompass[ɪn'kʌmpəs]v.包围，环绕，包含或包括某事物

encounter[ɪn'kaʊntə]vt.不期而遇

encourage[ɪn'kʌrɪdʒ]vt.鼓励

encrypt[ɪn'krɪpt]v.加密，将……译成密码

encryption[ɪn'krɪpʃn]n.加密，编密码

enforcement[ɪn'fɔ:smənt]n.执行，强制

engage[ɪn'geɪdʒ]vt.吸引住；

vi.与……建立密切关系；衔接；紧密结合

enhance[ɪn'hɑ:ns]vt.提高，增加；加强

enterprise['entəpraɪz]n.企业，事业

entertainment[ˌentə'teɪnmənt]n.娱乐

entity['entətɪ]n.实体

entrepreneurial[ˌɒntrəprə'nɜ:rɪəl]adj.企业家的，创业者的

environment[ɪn'vaɪrənmənt]n.环境，外界；工作平台

equipment[ɪ'kwɪpmənt]n.设备，装备

equivalent[ɪ'kwɪvələnt]adj.相等的，相当的，等效的

erase[ɪ'reɪz]vt.擦掉；抹去；清除

erroneous[ɪ'rəʊnɪəs]adj.错误的，不正确的

essential[ɪ'senʃl]adj.基本的；必要的；本质的

essentially[ɪ'senʃəlɪ]adv.本质上，根本上

establish[ɪ'stæblɪʃ]vt.建立，设立，安置，确定

establishment[ɪ'stæblɪʃmənt]n.设施，公司，确立，制定

etcetera[ˌet'setərə]n.其他，等等

Ethernet['i:θənet]n.以太网

ethical['eθɪkl]adj.道德的，伦理的

evaluation[ɪˌvælju'eɪʃn]n.估价，评价

exabyte['eksəbaɪt]n.艾字节

exact[ɪg'zækt]adj.精确的，准确的，原样的

exclusively[ɪk'sklu:sɪvlɪ]adv.专有地，排外地

execute['eksɪkju:t]vt.执行，实行

execution[ˌeksɪ'kju:ʃn]n.执行，实行，履行，执行

expandable[ɪk'spændəbl]adj.可扩展的，可扩大的

expansion[ɪk'spænʃn]n.扩充，扩展

expectation[ˌekspek'teɪʃn]n.期待；预期

expense[ɪk'spens]n.费用；消耗；vt.向……收取费用

expensive[ɪk'spensɪv]adj.昂贵的，花钱多的

experience[ɪk'spɪərɪəns]n.体验，经验；经历，阅历

vt.亲身参与，亲身经历；感受

experiment[ɪk'sperɪmənt]n.实验，试验；尝试；vi.做实验

expertise[ˌekspɜ:'ti:z]n.专门技能；专门知识；专门技术

explain[ɪk'spleɪn]v.解释，说明

exploitable[ɪks'plɔɪtəbl]adj.可开发的，可利用的

exploitation[ˌeksplɔɪ'teɪʃn]n.开发

explorable[ɪk'splɔ:rəbl]adj.可探索的

externally[ɪk'stɜ:nəlɪ]adv.外表上，外形上

extract[ɪk'strækt]vt.提取，析取

extraordinary[ɪk'strɔ:rdɪnərɪ]adj.非凡的；特别的

extremely[ɪk'stri:mlɪ]adv.极端地，非常地

F

fabricate['fæbrɪkeɪt]vt.编造，捏造

facility[fə'sɪlətɪ]n.设备，工具

factor['fæktə]n.因素，要素
failure['feɪljə]n.失败，失灵，故障，缺乏
fairly['feəlɪ]adv.适当地；清楚地；完全，简直
fascinating['fæsɪneɪtɪŋ]adj.迷人的，有极大吸引力的
fear[fɪə]n.害怕；可能性；
vt.害怕；为…忧虑（或担心、焦虑）；
vi.害怕；忧虑
feat[fi:t]n.技艺；功绩，壮举
feature['fi:tʃə]n.特征，特点
federate['fedəreɪt]v.（使）结成联盟
feed[fi:d]vt.馈送；向…提供
feedback['fi:dbæk]n.反馈，反应
feedback['fi:dbæk]n.反馈；反应
fiber-optic['faɪbə'ɔptɪk]n.光导纤维
field[fi:ld]n.域
filter['fɪltə]n.滤波器，筛选；vt.过滤
fingerprint['fɪŋgəprɪnt]n.指纹，手印；vt.采指纹
firewall['faɪəwɔ:l]n.防火墙
firmware['fɜ:mweə]n.固件，韧件（软件硬件相结合）
fixed[fɪkst]adj.固定的，确定的，准备好的
flexible['fleksəbl]adj.柔韧性，灵活的
float[fləʊt]n.浮点
forbidden[fə'bɪdn]adj.禁止的，严禁的
forecast['fɔ:kɑ:st]n.&vt.预测，预报
forefront['fɔ:frʌnt]n.前列；第一线；活动中心
formalize['fɔ:məlaɪz]vt.使正式，形式化
format['fɔ:mæt]n.格式；vt.格式化(磁盘)
formulate['fɔ:mjuleɪt]vt.规划，设计
forum['fɔ:rəm]n.论坛
forward['fɔ:wəd]vt.转发，转寄
foundation[faʊn'deɪʃn]n.基础，根本，建立，创立
fragment['frægmənt]n.碎片，断片，片段
frame[freɪm]n.帧，画面，框架
framework['freɪmwɜ:k]n.构架，框架，结构
fraudulent['frɔ:djələnt]adj.欺诈的，欺骗性的
free[fri:]adj.自由的，空闲的；vt.释放
freeware['fri:weə]n.免费软件
full-stack[fʊl-stæk]adj.全栈的；全能的；完整的
function['fʌŋkʃn]n.函数
functional['fʌŋkʃənl]adj.功能的；函数的
functionality[ˌfʌŋkʃə'nælətɪ]n.功能性
fundamentally[ˌfʌndə'mentəlɪ]adv.基础地，根本地
furnish['fɜ:nɪʃ]vt.供应，提供

G

gateway['geɪtweɪ]n.网关
gateway['geɪtweɪ]n.门；入口；途径
geek[gi:k]n.极客；对电脑痴迷的人
generalist['dʒenrəlɪst]n.通才，多面手
generate['dʒenəreɪt]vt.产生，发生
gift[gɪft]n.赠品，礼物
gigabit['ɡ ɪgəbɪt]n.吉字节，吉比特，千兆比特
glimpse[glɪmps]n.一瞥，一看；vi.瞥见
glove[glʌv]n.手套
grab[græb]v.抓住
gradually['grædʒuəlɪ]adv.逐步地，渐渐地
graphic['græfɪk]n.图形
grocery['grəʊsərɪ]n.食品杂货店，食品店，杂货铺
guarantee[ˌgærən'ti:]n.保证，担保；保证人；抵押品
vt.保证，担保

H

habit['hæbɪt]n.习惯，习性
hacker['hækə]n.黑客
halve[hɑ:v]vt.二等分，平分，分享，减半
handshake['hændʃeɪk]n.握手
haptic['hæptɪk]adj.触觉的
hardcopy['hɑ:d'kɒpɪ]n.硬拷贝
hardware['hɑ:dweə]n.硬件
hardware-dependent['hɑ:dweə-dɪ'pendənt]n.硬件相关；硬件依赖性
adj.硬件依赖的
headquarter['hed'kwɔ:tə]vi.设总部；
vt.将……的总部设在；把……放在总部里
headset['hedset]n.头套
healthcare['helθkeə]n.卫生保健
hide[haɪd]v.隐藏，掩藏，隐瞒
hierarchical[ˌhaɪə'rɑ:kɪkl]adj.分等级的，分层的
hierarchy['haɪərɑ:kɪ]n.分层，层次

hijack['haɪdʒæk]vt.劫持
honeynet['hʌnɪnet]n.蜜网
hop[hɒp]v.（鸟，蛙等）跳跃
horizontally[ˌhɒrɪ'zɒntəlɪ]adv.水平地，横地
host[həʊst]vt.托管；n.主机
hub[hʌb]n.集线器
huge[hju:dʒ]adj.巨大的，极大的
human-readable['hju:mən-'ri:dəbl]adj.人可读的
Hungary['hʌŋgərɪ]n.匈牙利
hybrid['haɪbrɪd]adj.混合的
hydrodynamics[ˌhaɪdrəʊdaɪ'næmɪks]n.流体力学，水动力学
hype[haɪp]n.大肆宣传，大做广告
hypermedia[ˌhaɪpə'mi:dɪə]n.超媒体
Hypertext['haɪpətekst]n.超文本
hypothesis[haɪ'pɒθəsɪs]n.假设，假说；前提

I

identification[aɪˌdentɪfɪ'keɪʃn]n.鉴定，识别
identifier[aɪ'dentɪfaɪə]n.标识符
identify[aɪ'dentɪfaɪ]vt.标识，识别，鉴别
identity[aɪ'dentətɪ]n.身份
ignore[ɪg'nɔ:]vt.不理睬，忽视
illicit[ɪ'lɪsɪt]adj.违法的
illusion[ɪ'lu:ʒn]n.错觉；幻想；假象
illustrate['ɪləstreɪt]vt.举例说明，图解，加插图于，阐明
imaginary[ɪ'mædʒɪnərɪ]adj.想象中的，假想的，虚构的
immediate[ɪ'mi:dɪət]adj.立即的；直接的
immense[ɪ'mens]adj.极广大的，无边的
immersive[ɪ'mɜ:sɪv]adj.拟真的，沉浸式的，浸入的
immigration[ˌɪmɪ'greɪʃn]n.移民总称；移居；移民入境
immunity[ɪ'mju:nətɪ]n.免疫性，免疫力
impede[ɪm'pi:d]vt.阻碍；妨碍；阻止
implement['ɪmplɪmənt]n.工具，器具；vt.贯彻，实现
implementation[ˌɪmplɪmen'teɪʃn]n.执行
impossible[ɪm'pɒsəbl]adj.不可能的，不会发生的，难以忍受的
incomplete[ˌɪnkəm'pli:t]adj.不完全的，不完善的
inconsistency[ˌɪnkən'sɪstənsɪ]n.矛盾
incorporate[ɪn'kɔ:pəreɪt]v.合并；adj.合并的
incorporating[ɪn'kɔ:pəreɪtɪŋ]v.融合，包含；使混合
increasingly[ɪn'kri:sɪŋlɪ]adv.越来越多地；日益，格外；越来越……
incremental[ˌɪnkrə'mentl]adj.增加的
indebtedness[ɪn'detɪdnəs]n.亏欠，债务
independent[ˌɪndɪ'pendənt]adj.独立的，不受约束的
independently[ˌɪndɪ'pendəntlɪ]adv.独立地
index['ɪndeks]n.索引；vt.编入索引中；vi.做索引
indicate['ɪndɪkeɪt]vt.表明，标示，指示
individual[ˌɪndɪ'vɪdʒuəl]n.个人，个体 adj.个别的，单独的，个人的
induction[ɪn'dʌkʃn]n.归纳
industrial[ɪn'dʌstrɪəl]adj.工业的，产业的
inexpensive[ˌɪnɪk'spensɪv]adj.便宜的，不贵重的
inextricably[ˌɪnɪk'strɪkəblɪ]adv.逃不掉地，解决不了地，解不开地
inflexibility[ɪnˌfleksə'bɪlətɪ]n.刚性，非灵活性
information[ˌɪnfə'meɪʃn]n.信息；数据；消息
infrastructure['ɪnfrəstrʌktʃə]n.基本设施
inherent[ɪn'hɪərənt]adj.固有的，内在的
inhibit[ɪn'hɪbɪt]v.抑制，约束
initialization[ɪˌnɪʃəlaɪ'zeɪʃn]n.设定初值，初始化
initially[ɪ'nɪʃəlɪ]adv.最初，开头
insider[ɪn'saɪdə]n.内部的人，知道内情的人，权威人士
inspect[ɪn'spekt]v.检查
instance['ɪnstəns]n.实例，建议，要求，情况，场合 vt.举……为例，获得例证
instantiate[ɪns'tænʃɪeɪt]vt.具体化，实例化
instigation[ˌɪnstɪ'geɪʃn]n.鼓动，教唆，煽动
institution[ˌɪnstɪ'tju:ʃn]n.公共机构，协会
instruct[ɪn'strʌkt]vt.指示，通知；命令
instrumental[ˌɪnstrə'mentl]adj.有帮助的，起作用的；乐器的；仪器的
intact[ɪn'tækt]adj.完整无缺的
integer['ɪntɪdʒə]n.整数
integrate['ɪntɪgreɪt]vt.使成整体，使一体化，求……的积分
integrated['ɪntɪgreɪtɪd]adj.集成的，综合的，完整的
integration[ˌɪntɪ'greɪʃn]n.结合；整合；一体化
integrative['ɪntɪgreɪtɪv]adj.综合的，一体化的

intention[ɪn'tenʃn]*n.*意图，目的；意向
interact[ˌɪntər'ækt]*vi.*互相作用，互相影响
interaction[ˌɪntər'ækʃn]*n.*交互作用
interactive[ˌɪntər'æktɪv]*adj.*交互式的
intercept[ˌɪntə'sept]*vt.*中途阻止，截取
interconnect[ˌɪntəkə'nekt]*vi.*互相连接，互相联系
*vt.*使互相连接；使互相联系
interface['ɪntəfeɪs]*n.*接口，界面
intermediate[ˌɪntə'mi:dɪət]*adj.*中间的；*n.*媒介
internationally[ˌɪntə'næʃnəlɪ]*adv.* 国际性地，国际上地，国际地
Internet[ɪn'tənet]*n.*因特网
Internet['ɪntənet]*n.*互联网
internetwork[ɪntə:r'netwɜ:k]*n.*网间
interoperability['ɪntərɒpərə'bɪlətɪ]*n.*互通性，互用性，互操作性
interpret[ɪn'tɜ:prət]*vt.*解释；*vi.*做解释
interpreter[ɪn'tɜ:prtə]*n.*解释程序，解释器
interrelationship[ˌɪntərɪ'leɪʃnʃɪp]*n.*相互关系，相互关联
interrupt[ˌɪntə'rʌpt]*vt.& n.*中断
intervention[ˌɪntə'venʃn]*n.*介入，干涉，干预
intrinsic[ɪn'trɪnsɪk]*adj.*固有的，内在的，本质的
intrusion[ɪn'tru:ʒn]*n.*闯入，侵扰
intuitive[ɪn'tju:ɪtɪv]*adj.*直觉的；直观的
invaluable[ɪn'væljuəbl]*adj.*无价的，价值无法衡量的
invariable[ɪn'veəriəbl]*adj.*不变的，常量
investigation[ɪnˌvestɪ'geɪʃn]*n.*调查，研究
irrecoverable[ˌɪrɪ'kʌvərəbl]*adj.*无可挽救的；不可弥补
iterative['ɪtərətɪv]*adj.*重复的，反复的，迭代的

J

join[dʒɔɪn]*vt.*连接，结合，参加，加入
joystick['dʒɔɪstɪk]*n.*操纵杆
junk[dʒʌŋk]*vt.*丢弃，废弃 *n.*废品；假货

K

keyboard['ki:bɔ:d]*n.*键盘
knowledge['nɒlɪdʒ]*n.*知识；了解，理解

L

label['leɪbl]*vt.*贴标签于，指……为，分类，标注
*n.*标签，标志
landmark['lændmɑ:k]*n.*界标；里程碑；纪念碑
latency['leɪtənsɪ]*n.*反应时间；潜伏，潜在，潜伏物
lawmaker['lɔ:meɪkə]*n.*立法者
layered['leɪəd]*adj.*分层的
legitimate[lɪ'dʒɪtɪmət]*adj.*合法的，合理的
*v.*合法
lens[lenz]*n.*镜头；透镜
liability[ˌlaɪə'bɪlətɪ]*n.*责任，义务，倾向，债务
liaison[li'eɪzn]*n.*联络
library['laɪbrərɪ]*n.*库
linear['lɪnɪə]*adj.*直线的；线性的
link[lɪŋk]*n.&v.*链接
linkage['lɪŋkɪdʒ]*n.*连接，联结
liteware[laɪtweə]*n.*精简版软件
local['ləʊkl]*adj.*本地的
location[ləʊ'keɪʃn]*n.*位置，场所
log[lɒg]*n.*日志
logical['lɒdʒɪkl]*adj.*逻辑的；符合逻辑的
loop[lu:p]*n.*循环

M

machine-readable[mə'ʃi:n-'ri:dəbl]*adj.*计算机可读的，可用计算机处理的
macro['mækrəʊ]*n.*宏
mail-order[ˌmeɪl-'ɔ:də]*adj.*邮购的
mainframe['meɪnfreɪm]*n.*主机，大型计算机
mainstream['meɪnstri:m]*n.*主流
maintain[meɪn'teɪn]*vt.*维持，维修，维护
maintainable[meɪn'teɪnəbl]*adj.*可维护的
maintenance['meɪntənəns]*n.*维护，保持
malicious[mə'lɪʃəs]*adj.*怀恶意的，恶毒的
malware['mælweə]*n.*恶意软件，流氓软件
manipulate[mə'nɪpjuleɪt]*vt.*操作，使用
manually['mænjuəlɪ]*adv.*用手，用动的
map[mæp]*n.*映射
masquerade[ˌmæskə'reɪd]*v.*化装；伪装
mass[mæs]*n.*大量，大多；*v.*（使）集中，聚集
*adj.*大规模的；整个的；集中的
match[mætʃ]*n.&v.*匹配
matrix['meɪtrɪks]*n.*矩阵

mechanism['mekənɪzəm]n.机制
media['mi:dɪə]n.媒体，介质
memory['memərɪ]n.存储器，内存
mentally['mentəlɪ]adv.心理上，精神上；智力上
menu['menju:]n.菜单
merchant['mɜ:tʃənt]n.商人，批发商，贸易商，店主
adj.商业的，商人的
merge[mɜ:dʒ]v.合并，并入，结合
message['mesɪdʒ]n.消息，通信，讯息；vt.通知
metadata['metədeɪtə]n.元数据
metaphor['metəfə]n.隐喻，暗喻，比喻
meteorology[ˌmi:tɪə'rɒlədʒɪ]n.气象学；气象状态
method['meθəd]n.方法
methodology[ˌmeθə'dɒlədʒɪ]n.方法学，方法论
microcode['maɪkrəʊkəʊd]n.微代码，微码
microprocessor[ˌmaɪkrəʊ'prəʊsesə]n.微处理器
middleware['mɪdlweə]n.中间件
minimize['mɪnɪmaɪz]vt.把……减至最低数量（程度），最小化
misuse[ˌmɪs'ju:z]v.& n.误用，错用，滥用
mixture['mɪkstʃə]n.混合，混合物
mobile['məʊbaɪl]adj.可移动的
mockup['mɒkʌp]n.制造模型；制造样机；实体模型
modem['məʊdem]n.调制解调器
modification[ˌmɒdɪfɪ'keɪʃn]n.更改，修改，修正
modify['mɒdɪfaɪ]vt.更改，修改
modular['mɒdjələ]adj.模块化的
modularity[mɒdju'lærɪtɪ]n.模块性
modulate['mɒdjuleɪt]vt.（信号）调制
module['mɒdju:l]n.模块
moment['məʊmənt]n.瞬间，片刻
motherboard['mʌðəbɔ:d]n.主板，母板
motivate['məʊtɪveɪt]v.激发，促进
mouse[maʊs]n.鼠标器
multi-core['mʌlti-kɔ:]adj.多核的
multimedia[ˌmʌltɪ'mi:dɪə]n.多媒体；adj.多媒体的
multiple['mʌltɪpl]adj.多重的；多个的；多功能的
multiplicity[ˌmʌltɪ'plɪsətɪ]n.多样性
multiport['mʌltɪpɔ:t]adj.多端口的
multiprocessing['mʌltɪˌprəʊsesɪŋ]n.多重处理，多处理（技术）
multiprogramming['mʌltɪˌprəʊgræmɪŋ]n.多道程序设计，多程序设计
multistation[mʌltɪs'teɪʃn]n.多站
multi-tasking['mʌltɪ-'tɑ:skɪŋ]n.多任务
multitudinousness[ˌmʌltɪ'tju:dɪnəsnɪs]n.非常多；众多的
multi-user['mʌltɪ-'ju:zə]n.多用户
mutation[mju:'teɪʃn]n.变化，转变，(生物物种的)突变

N

naivety[naɪ'i:vətɪ]n.天真烂漫，单纯
narrow['nærəʊ]adj.狭窄的，精密的，严密的，有限的
narrow['nærəʊ]adj.狭隘的，狭窄的
neglected[nɪ'glektɪd]adj.被忽视的
network['netwɜ:k]n.网络
neutral['nju:trəl]adj.中立的
niche[nɪtʃ]n.小生态环境，细分市场，商机
node[nəʊd]n.节点
nonimmersive[nɒnɪ'mɜ:sɪv]adj.非拟真的，非沉浸式的，非浸入的
nonintuitive[nɒnɪn'tju:ɪtɪv]adj.非直觉的
non-linear[nɒn-'lɪnɪə]adj.非线性的
nonvolatile['nɒn'vɒlətaɪl]adj.非易失性的
normalization[ˌnɔ:məlaɪ'zeɪʃn]n.归一化，正规化，标准化，规格化，规范化
nosy['nəʊzɪ]adj.好管闲事的，爱追问的；n.好管闲事的人
notation[nəʊ'teɪʃn]n.标记符号，表示法
notion['nəʊʃn]n.概念，观念，想法
notoriously[nəʊ'tɔ:rɪəslɪ]adj.声名狼藉地，臭名昭著地，众所周知地
novelty['nɒvltɪ]n.新奇，新奇的事物
numeric[nju:'merɪk]adj.数字的

O

obey[ə'beɪ]v.服从，听从
obituary[ə'bɪtʃuərɪ]n.讣告
object-oriented['ɒbdʒektɔ:rɪəntɪd]adj.面向对象的
observation[ˌɒbzə'veɪʃn]n.观察，观察力
obstruction[əb'strʌkʃn]n.阻塞，阻碍，受阻

obvious['ɒbvɪəs]*adj.*明显的；显著的
occupation[ɒkju:'peɪʃn]*n.*职业，工作
off-premise['ɔ:f'premɪs]*adj.*外部部署的
offsite[ɒf'saɪt]*adj.*异地的
on-line['ɒn'laɪn]*n.*联机，在线式
on-premise['ɒn'premɪs]*adj.*本地部署的
ontology[ɒn'tɒlədʒɪ]*n.*本体，存在；实体论
opaque[əʊ'peɪk]*adj.*不透明的；含糊的；*n.*不透明
operation[ˌɒpə'reɪʃn]*n.*运算；操作，经营
opponent[ə'pəʊnənt]*n.*对手
optimal['ɒptɪməl]*adj.*最佳的，最优的；最理想的
optimization[ˌɒptɪmaɪ'zeɪʃən]*n.*最佳化，最优化
option['ɒpʃn]*n.*选项
orchestrate['ɔ:kɪstreɪt]*vt.*精心策划
organ['ɔ:gən]*n.*器官；元件
orientation[ˌɔ:rɪən'teɪʃn]*n.*方向，定位，取向
originate[ə'rɪdʒɪneɪt]*vi.*起源，发生
outage['aʊtɪdʒ]*n.*断供
outdated[ˌaʊt'deɪtɪd]*adj.*过时的，不流行的
outpace[ˌaʊt'peɪs]*vt.*超过……速度，赶过
outsourceable['aʊtsɔ:səbl]*adj.*可外界供应的，可外包的
outsourcing['aʊtsɔ:sɪŋ]*n.*外包，外购
overlap[ˌəʊvə'læp]*v.*（与……）交叠，（与……）重叠
oversee[ˌəʊvə'si:]*v.*监视，检查
oversimplification['əʊvəˌsɪmplɪfɪ'keɪʃn]*n.*过度简化（的事物）
overwhelming[ˌəʊvə'welmɪŋ]*adj.*势不可挡的，压倒一切的

P

package['pækɪdʒ]*n.*包裹，包
packet['pækɪt]*n.*包，信息包；*v.*包装
paging['peɪdʒɪŋ]*n.*分页
parallel['pærəlel]*adj.*并行的，相同的
parallelism['pærəlelɪzəm]*n.*平行，对应，类似
paravirtualization['pærəvɜ:tʃʊəlaɪ'zeɪʃn]*n.*准虚拟化，半虚拟化
parse[pɑ:z]*vt.*解析
partially['pɑ:ʃəlɪ]*adv.*部分地
participate[pɑ:'tɪsɪpeɪt]*vi.*参与，参加，分享，分担
particular[pə'tɪkjələ]*adj.*特别的；详细的；独有的 *n.*特色，特点
particularly[pə'tɪkjələlɪ]*adv.*独特地，显著地
passcode[pɑ:skəʊd]*n.*密码
passive['pæsɪv]*adj.*被动的
password['pɑ:swɜ:d]*n.*密码，口令
path[pɑ:θ]*n.*路经
path[pɑ:θ]*n.*路径
pattern['pætn]*n.*式样，模式
payment['peɪmənt]*n.*付款，支付
pedestrian[pə'destrɪən]*n.*步行者，行人 *adj.*徒步的
peer[pɪə]*n.*同等的人；同辈；同龄人
perceive[pə'si:v]*vt.*察觉；理解；发觉；意识到
perception[pə'sepʃn]*n.*知觉，觉察
perform[pə'fɔ:m]*v.*执行
peripheral[pə'rɪfərəl]*adj.*外围的；次要的；*n.*外部设备
permanent['pɜ:mənənt]*adj.*永久（性）的，不变的，持久的
permanently['pɜ:mənəntlɪ]*adv.*永存地，不变地
permit[pə'mɪt]*v.*许可，准许；*n.*许可，准许；许可证
persistence[pə'sɪstəns]*n.*坚持不懈；执意；持续
perspective[pə'spektɪv]*n.*观点，看法，观点，观察
petabyte['petəbait]*n.*拍字节
phase[feɪz]*n.*阶段
photocell['fəʊtəʊsel]*n.*光电池
piece[pi:s]*n.*块，件，片
piezoelectric[paɪˌi:zəʊɪ'lektrɪk]*adj.*压电的
pilot['paɪlət]*n.*引航员；向导；*vt.*驾驶
pinpoint['pɪnpɔɪnt]*n.*精确；*adj.*极微小的；*v.*查明
pitfall['pɪtfɔ:l]*n.*潜在的困难；陷阱，圈套，诱惑
platform['plætfɔ:m]*n.*平台
plausible['plɔ:zəbl]*adj.*貌似有道理的
pocket['pɒkɪt]*n.*口袋
pointer['pɔɪntə]*n.*指针
policy['pɒləsɪ]*n.*政策，方针
pool[pu:l]*n.*池
poorly['pʊəlɪ]*adj.*恶劣的
popularity[ˌpɒpju'lærətɪ]*n.*普及，流行
port[pɔ:t]*n.*端口

portability[ˌpɔ:təˈbɪlətɪ]n.可移植性，可携带，轻便
position[pəˈzɪʃn]n.位置；vt.安置，决定……的位置
possess[pəˈzes]vt.拥有；掌握，懂得
possibility[ˌpɒsəˈbɪlətɪ]n.可能性
potential[pəˈtenʃl]adj.潜在的，可能的
potentially[pəˈtenʃəlɪ]adv.潜在地
potentiometer[pəˌtenʃɪˈɒmɪtə]n.电位计，电势计
powerful[ˈpaʊəfl]adj.强大的，有力的
practice[ˈpræktɪs]n.实行，实践，实际，惯例，习惯
practitioner[prækˈtɪʃənə]n.从业者
prebuilt[preˈbɪlt]adj.预建的，预制的
precaution[prɪˈkɔ:ʃn]n.预防，防备，警惕；预防措施
vt.使提防
predictable[prɪˈdɪktəbl]adj.可预言的
prediction[prɪˈdɪkʃn]n.预测，预报；预言
predictive[prɪˈdɪktɪv]adj.预言性的
preemptive[prɪˈemptɪv]adj.抢占式的
prehistory[ˌpri:ˈhɪstrɪ]n.史前时期
preload[ˌpri:ˈləʊd]n.&v.预载，预先装载
preparation[ˌprepəˈreɪʃn]n.准备，预备
prescribe[prɪˈskraɪb]v.指示，规定
prescription[prɪˈskrɪpʃn]n.处方，药方
presentation[ˌpreznˈteɪʃn]n.表示，表达，介绍，陈述
preserve[prɪˈzɜ:v]vt.保存，保持，保护
prevent[prɪˈvent]v.防止，预防
primitive[ˈprɪmətɪv]adj.原始的，简单的
principal[ˈprɪnsəpl]adj.主要的，首要的
principle[ˈprɪnsəpl]n.法则，原则，原理
printer[ˈprɪntə]n.打印机
priority[praɪˈɒrətɪ]n.优先，优先权
privilege[ˈprɪvɪlədʒ]n.特权，特别待遇
vt.给予……特权，特免
pro[prəʊ]adv.正面地；n.赞成的意见；赞成票，优点
proactive[ˌprəʊˈæktɪv]adj.积极的，主动地
probe[prəʊb]v.探查，探测
procedural[prəˈsi:dʒərəl]adj.过程化的，程序上的
procedure[prəˈsi:dʒə]n.过程
process[ˈprəʊses]vt.加工；处理；n.过程
product[ˈprɒdʌkt]n.产品，产物，乘积
productivity[ˌprɒdʌkˈtɪvətɪ]n.生产率，生产力
profess[prəˈfes]vt.声称；宣称；公开表明；信奉
profit[ˈprɒfɪt]n.利润，益处，得益
vi.得益，利用；vt.有益于，有利于
profitable[ˈprɒfɪtəbl]adj.有利可图的
program[ˈprəʊgræm]n.程序；v.编写程序
programmable[ˈprəʊgræməbl]adj.可编程的，可设计的
programmer[ˈprəʊgræmə]n.程序设计员
programming[ˈprəʊgræmɪŋ]n.编程
progress[ˈprəʊgres]n.进程，发展
promise[ˈprɒmɪs]vt.& n.允诺，答应，许诺
prompt[prɒmpt]v.提示
proper[ˈprɒpə]adj.适当的，正确的，固有的，特有的
properly[ˈprɒpəlɪ]adv.适当地，完全地
property[ˈprɒpətɪ]n.性质，特性
proprietary[prəˈpraɪətrɪ]adj.私有的，私有财产的
protocol[ˈprəʊtəkɒl]n.协议
prototype[ˈprəʊtətaɪp]n.原型
provide[prəˈvaɪd]v.供应，供给，准备
provision[prəˈvɪʒn]n.供应，预备，规定
psychology[saɪˈkɒlədʒɪ]n.心理学；心理特点；心理状态
publish[ˈpʌblɪʃ]vt.公开；发表；出版；发行
pulse[pʌls]n.脉冲
puzzle[ˈpʌzl]vt.使迷惑，使难解；n.难题
pyramid[ˈpɪrəmɪd]n.金字塔；角堆体
v.（使）成金字塔状，（使）渐增，（使）上涨

Q

quadrant[ˈkwɒdrənt]n.象限
quality[ˈkwɒlətɪ]n.质量，品质
quantity[ˈkwɒntətɪ]n.量，数量
query[ˈkwɪərɪ]n.&v.查询

R

random[ˈrændəm]adj.任意的；随机的
n.随意；偶然的行动
rating[ˈreɪtɪŋ]n.评估，评价；v.估价；定级
reach[ri:tʃ]vt.到达，达到，伸出，延伸
reactive[rɪˈæktɪv]adj.反应的
readability[ˌri:dəˈbɪlətɪ]n.可读性

readable['ri:dəbl]adj.可读的，易读的；易懂的
realistic[ˌri:ə'lɪstɪk]adj.逼真的；栩栩如生的
reallocation[ˌri:ˌælə'keɪʃn]vt.再分配
realm[relm]n.领域，范围
reasoning['ri:zənɪŋ]n.推理，论证；v.推理，思考；争辩；说服；adj.推理的
reassemble[ˌri:ə'sembl]vt.重新装配，重新集合
recipe['resəpɪ]n.处方
recognition[ˌrekəg'nɪʃn]n.识别
recompile[rɪkəm'pɑɪl]vt.重新编译
reconnaissance[rɪ'kɒnɪsns]n.侦察，搜索
reconstruction[ˌri:kən'strʌkʃn]n.复原物；重建物；重建；再现
record['rekɔ:d]n.记录
recursive[rɪ'kɜ:sɪv]adj.回归的，递归的
redesign[ˌri:dɪ'zaɪn]v.重新设计；n.重新设计，新设计
redistribute[ˌri:dɪ'strɪbju:t]vt.重现发布；重新分配
redundancy[rɪ'dʌndənsi]n.冗余
refactoring[rɪ'fæktərɪŋ]n.重构
reference['refrəns]n.提及，涉及，参考
reflect[rɪ'flekt]v.反射
regenerate[rɪ'dʒenəreɪt]vt.使新生，重建；vi.新生，再生
adj.新生的，更新的
register['redʒɪstə]n.寄存器
regression[rɪ'greʃn]n.（统计学）回归
regular['regjulə]adj.规则的，有秩序的，合格的，定期的
regulate['regjuleɪt]vt.控制，管理；调节，调整；校准
regulation[ˌregju'leɪʃn]n.规章，规则；adj.规定的
reject[rɪ'dʒekt]vt.拒绝，抵制，否决，丢弃
relationship[rɪ'leɪʃnʃɪp]n.关联，关系
relay['ri:leɪ]n.继电器
vt.转发，转播，传达
reliability[rɪˌlaɪə'bɪlətɪ]n.可靠性
remainder[rɪ'meɪndə]n.剩余物
remote[rɪ'məʊt]adj.远程的；n.远程操作
remove[rɪ'mu:v]vt.删除，移去
repeatable[rɪ'pi:təbl]adj.可重复的
repeatedly[rɪ'pi:tɪdlɪ]adv.反复地，重复地
repeater[rɪ'pi:tə]n.转发器；中继器
repetitive[rɪ'petɪtɪv]adj.重复的，啰唆的
replica['replɪkə]n.复制品
representation[ˌreprɪzen'teɪʃn]n.表示法，表现，代表
reprogram[rɪ'prəʊgræm]v.重新编程，程序重调
reputable['repjətəbl]adj.声誉好的，有声望的，有好评的
resource[rɪ'sɔ:s]n.资源
responsibility[rɪˌspɒnsə'bɪlətɪ]n.责任，职责
responsible[rɪ'spɒnsəbl]adj.负责的，可靠的，可依赖的
restriction[rɪ'strɪkʃn]n.限制，约束
retail['ri:teɪl]n.零售；adj.零售的；v.零售
retain[rɪ'teɪn]vt.保持，保留
retrospective[ˌretrə'spektɪv]adj.回顾的
reusable[ˌri:'ju:zəbl]adj.可再用的，可重复使用的
reuse[ˌri:'ju:z]vt.重用，复用
revenue['revənju:]n.收入，税收
revise[rɪ'vaɪz]vt.修订，校订，修改
reward[rɪ'wɔ:d]n.奖赏；报酬；赏金；酬金
vt.奖赏；酬谢
rework[ˌri:'wɜ:k]vt.重做，改写，重写
rewritable[ˌri:'raɪtəbl]adj.可重写的，可复写的
risk[rɪsk]vt.冒……的危险；n.冒险，风险
rival['raɪvl]n.竞争者，对手；v.竞争，对抗，相匹敌
robot['rəʊbɒt]n.机器人；遥控装置
router['ru:tə]n.路由器
row[rəʊ]n.行，排
rule[ru:l]n.规则，规定；统治，支配；v.控制，支配

S

safeguard['seɪfgɑ:d]vt.维护，保护，捍卫
n.安全装置，安全措施
scalability[skeɪlə'bɪlɪtɪ]n.可扩展性，可量测性
scalable['skeɪləbl]adj.可扩展的，可升级的
scan[skæn]v.& n.扫描
scanner['skænə]n.扫描器；扫描设备
scenario[sə'nɑ:rɪəʊ]n.设想；可能发生的情况；剧情梗概
scene[si:n]n.场面，现场
schedule['ʃedju:l]n.预定计划；时刻表，进度表
vt.排定，安排
schema['ski:mə]n.模式，大纲，模型

scour['skaʊə]v.四处搜集，冲洗，擦亮
screen[skri:n]n.屏幕
script[skrɪpt]n.脚本
scrum[skrʌm]n.橄榄球里的争球
seabed['si:bed]n.海底，海床
seasonal['si:zənl]adj.季节性的
security[sɪ'kjʊərətɪ]n.安全
seemingly['si:mɪŋlɪ]adv.表面上地
segment['segmənt]n.段，节，片断；v.分割
segmentation[ˌsegmen'teɪʃn]n.分割；分段；分节
selective[sɪ'lektɪv]adj.选择的，选择性的
self-awareness[self-ə'weənɪs]n.自我意识
self-correction[ˌselfkə'rekʃn]n.自校正；自我纠错；自我改正
semi-standardized['semɪ-'stændədaɪzd]adj.半标准化的
semi-structured['semɪ-'strʌktʃəd]adj.半结构化的
semi-transparent['semɪtrænsp'ærənt]adj.半透明的
send[send]vt.送，寄，发送
sender['sendə]n.寄件人，发送人
sensor['sensə]n.传感器
sensory['sensərɪ]adj.传递感觉的
sentient['sentɪənt]adj.有感觉能力的，有知觉力的
separate['sepərət]['sepəreɪt]adj.分开的，分离的，单独的 v.分开，隔离，分散
separated['sepəreɪtɪd]adj.分隔的，分开的
separately['seprətlɪ]adv.个别地，分离地
separating['sepəreɪtɪŋ]adj.分开的，分离的
sequence['si:kwəns]n.次序，顺序，序列
serial['sɪərɪəl]adj.串行的
serve[sɜ:v]v.服务，供应，适合
server['sɜ:və]n.服务器
session['seʃn]n.会话，对话
set[set]n.集合；vt.设置
settlement['setlmənt]n.定居点
share[ʃeə]v.共享，分享
shareware['ʃeəweə]n.共享软件
sheer[ʃɪə]adj.全然的，纯粹的，绝对的，彻底的
shift[ʃɪft]n.&v.移动，移位，变化
ship[ʃɪp]v.运输，载运
sift[sɪft]vt.筛分，精选；审查；vi.筛；细查
signal['sɪgnəl]n.信号；adj.信号的；v.发信号，用信号通知
signaling['sɪgnəlɪŋ]n.发信号
signature['sɪgnətʃə]n.签名；署名；识别标志
significance[sɪg'nɪfikəns]n.意义，意思；重要性
significant[sɪg'nɪfikənt]adj.有意义的，重大的，重要的
significantly[sɪg'nɪfikəntlɪ]adv.意味深长地；值得注目地
similarity[ˌsɪmə'lærətɪ]n.相像性，相仿性，类似性
simulate['sɪmjuleɪt]vt.模拟，模仿
simulating['sɪmjʊleɪtɪŋ]n.模拟 v.模仿；（用计算机或模型等）模拟
simulation[ˌsɪmju'leɪʃn]n.仿真，模拟
simulator['sɪmjuleɪtə]n.模拟装置，模拟器
simultaneously[ˌsɪml'teɪnɪəslɪ]adv.同时地；一齐
single-tasking['sɪŋgl-'tɑ:skɪŋ]n.单任务
single-user['sɪŋgl-'ju:zə]n.单用户
site[saɪt]n.网站
situated['sɪtjueɪtɪd]adj.位于，被置于境遇
situation[ˌsɪtju'eɪʃn]n.（人的）情况；局面，形势，处境；位置
smartphone[smɑ:tfəʊn]n.智能手机
software['sɒftweə]n.软件
sophisticated[sə'fɪstɪkeɪtɪd]adj.复杂的；精致的；富有经验的
sorting['sɔ:tɪŋ]n.排序
spawn[spɔ:n]vt.大量生产
spectrum['spektrəm]n.光谱；波谱；范围；系列
spiral['spaɪrəl]adj.螺旋形的；n.螺旋
sponsorship['spɒnsəʃɪp]n.发起，主办；倡议
spoof[spu:f]v.哄骗，欺骗
spot[spɒt]vt.认出，发现
spreadsheet['spredʃi:t]n.电子制表软件，电子数据表
spyware['spaɪweə]n.间谍软件
stability[stə'bɪlətɪ]n.稳定（性）；稳固
stack[stæk]n.堆，堆栈；v.堆叠
stage[steɪdʒ]n.发展的进程，阶段或时期
stalker['stɔ:kə]n.跟踪者
standpoint['stændpɔɪnt]n.立场，观点

start-up[stɑ:t-ʌp]n.启动
stateless['steɪtləs]a.无状态的
statement['steɪtmənt]n.语句
static['stætɪk]adj.静止的；不变的
station['steɪʃn]n.站点
statistical[stə'tɪstɪkl]adj.统计的，统计学的
statistics[stə'tɪstɪks]n.统计学；统计表
status['steɪtəs]n.状态
stereoscopic[ˌsterɪə'skɒpɪk]adj.有立体感的
storage['stɔ:rɪdʒ]n.存储
store[stɔ:]v.（在计算机里）存储；贮存
straightforward[ˌstreɪt'fɔ:wəd]adj.直截了当的；坦率的；明确的
adv.直截了当地；坦率地
strategy['strætədʒɪ]n.策略
streamline['stri:mlaɪn]n.流线型
strengthen['streŋθn]v.加强，巩固
stretch[stretʃ]v.伸展；延伸
strict[strɪkt]adj.严格的；精确的；绝对的
strive[straɪv]v.努力，奋斗
sublayer['sʌb'leɪə]n.子层，下层
submodule[sʌb'mɒdju:l]n.子模块
sub-routine['sʌbru:ti:n]n.子例程
subsection['sʌbsekʃn]n.分部，分段，小部分，小单位
subsequent['sʌbsɪkwənt]adj.后来的
subset['sʌbset]n.子集
substantial[səb'stænʃl]adj.实质的，真实的，充实的
substitute['sʌbstɪtju:t]v.代替，替换，代用
subsystem[sʌb'sɪstəm]n.子系统，分系统
subtle['sʌtl]adj.微妙的；敏感的
successive[sək'sesɪv]adj.继承的，连续的
suit[su:t]v.合适，适合
suitable['su:təbl]adj.适当的，相配的
supercomputer['su:pəkəmpju:tə]n.超级计算机，巨型计算机
superscalar['su:pəskeɪlə]n.超标量结构
supervisor['su:pəvaɪzə]n.管理者；监督者；指导者
supplier[sə'plaɪə]n.供应商，厂商，供给者
surrounding[sə'raʊndɪŋ]n.环境；adj.周围的
surveillance[sɜ:'veɪləns]n.监视，监督
suspend[sə'spend]v.挂起，暂停
sweep[swi:p]v.扫过，掠过
switch[swɪtʃ]n.交换机
synchronization[ˌsɪŋkrənaɪ'zeɪʃn]n.同步
synergism['sɪnədʒɪzəm]n.合作
syntactically[sɪn'tæktɪklɪ]adv.依照句法地，在语句构成上
systematic[ˌsɪstə'mætɪk]adj.系统的，体系的
systematize['sɪstəmətaɪz]v.系统化

T

table['teɪbl]n.表，表格；vt.制表
tamper['tæmpə]vi.篡改
tandem['tændəm]adv.一个跟着一个地
tangible['tændʒəbl]adj.可触知的；确实的，真实的；实际的
task[tɑ:sk]n.工作，任务；作业
taste[teɪst]n.味道，滋味；味觉
taxonomy[tæk'sɒnəmɪ]n.分类学，分类系统
tenancy['tenənsɪ]n.租用，租赁；租期
terminator['tɜ:mɪneɪtə]n.终接器
textual['tekstʃuəl]adj.本文的
tidbit['tɪdbɪt]n.珍品，珍闻
tilt[tɪlt]v.倾斜；n.倾斜，歪斜
time-consuming[taɪm-kən'sju:mɪŋ]adj.耗费时间的，旷日持久的
time-dependent[taɪm-dɪ'pendənt]n.时间相关；时间依赖性
adj.时间依赖的
time-sharing['taɪm'ʃeərɪŋ]n.分时
token['təʊkən]n.令牌
tolerance['tɒlərəns]n.容错
topology[tə'pɒlədʒɪ]n.拓扑
touchscreen['tʌtʃskri:n]n.触摸屏
traditional[trə'dɪʃənl]adj.传统的，惯例的
traffic['træfɪk]n.流量，通信量
train[treɪn]v.训练；教育；培养
transceiver[træn'si:və]n.收发器
transform[træns'fɔ:m]vt.改变；改观；变换；vi.改变
transit['trænzɪt]vt.传输

translate[træns'leɪt]vt.翻译，解释，转化
transmit[træns'mɪt]vt.传输，传播
transparent[træns'pærənt]adj.透明的，显然的，明晰的
transparently[træns'pærəntlɪ]adv.显然地，易察觉地
transport['trænspɔ:t]n.传送器，运输；vt.传送，运输
transportation[ˌtrænspɔ:'teɪʃn]n.运输，运送
treatment['tri:tmənt]n.处理，治疗
tremendous[trə'mendəs]adj.极大的，巨大的；极好的
trigger['trɪgə]vt.引发，触发
trivial['trɪvɪəl]adj.普通的，平凡的，价值不高的，微不足道的
Trojan['trəʊdʒən]n.特洛伊
troubleshoot['trʌblʃu:t]v.故障排除,故障检测
trunk[trʌŋk]n.干线，中继线
tutelage['tju:təlɪdʒ]n.教导；指导

U

unauthorized[ʌn'ɔ:θəraɪzd]adj.未被授权的，未经认可的
unavailable[ˌʌnə'veɪləbl]adj.不可用的，无效的
unavoidable[ˌʌnə'vɔɪdəbl]adj.不可避免的，不得已的
unclear[ˌʌn'klɪə]adj.不清楚的，不明白的，含糊不清
uncover[ʌn'kʌvə]vt.揭开，揭露，揭示
underpin[ˌʌndə'pɪn]vt.加固，支撑
unexpected[ˌʌnɪk'spektɪd]adj.意外的；忽然的；突然的
unfamiliar[ˌʌnfə'mɪlɪə]adj.不熟悉的；不常见的；陌生的；没有经验的
unfortunately[ʌn'fɔ:tʃənətlɪ]adv.不幸地
union['ju:nɪən]n.联合，合并，结合
unique[ju:'ni:k]adj.唯一的，独特的
uniquely[jʊ'ni:klɪ]adv.独特地，唯一地
unmanage[ʌn'mænɪdʒ]vi.不处理，应付过去
vt.不管理，不控制，不操纵，不维持
unmet[ˌʌn'met]adj.未满足的，未相遇的，未应付的
unnoticed[ˌʌn'nəʊtɪst]adj.不引人注意的，被忽视的
unprocessed[ʌn'prəʊsest]adj.未处理的
unrealistic[ˌʌnrɪə'lɪstɪk]adj.不切实际的；不现实的；空想的
unrest[ʌn'rest]n.不安的状态，动荡的局面
unrestricted[ˌʌnrɪ'strɪktɪd]adj.不受限制的
unruly[ʌn'ru:lɪ]adj.不受拘束的，不守规矩的
unstructured[ʌn'strʌktʃəd]adj.非结构化的，无结构的
unsupervised[ˌʌn'sju:pəvaɪzd]adj.无人监督的，无人管理的
untapped[ˌʌn'tæpt]adj.未使用的
update[ʌp'deɪt]['ʌpdeɪt]vt.更新，使现代化
n.更新
usability[ˌju:zə'bɪlətɪ]n.可用性
username['ju:zəneɪm]n.用户名
utility[ju:'tɪlətɪ]n.实用（程序），效用，有用

V

valid['vælɪd]adj.有效的，有根据的，正当的，正确的
validate['vælɪdeɪt]vt.确认；证实
validation[ˌvælɪ'deɪʃn]n.确认，有效性
valuable['væljuəbl]adj.有价值的
valuation[ˌvælju'eɪʃn]n.估价，评价，计算
variable['veərɪəbl]n.变量
variation[ˌveəri'eɪʃn]n.变种，变更，变异，变化
vector['vektə]n.向量，矢量
vehicle['vi:əkl]n.车辆；交通工具
vein[veɪn]n.矿脉，纹理
vendor['vendə]n.卖主
verification[ˌverɪfɪ'keɪʃn]n.确认
versatility[ˌvɜ:sə'tɪlətɪ]n.多用途
vertical['vɜ:tɪkl]adj.垂直的，竖立的；n.垂直线，垂直面
virtual['vɜ:tʃuəl]adj.（计算机）虚拟的；实质上的，事实上的
virtualized['vɜ:tʃuəlaɪzd]adj.虚拟化的
vision['vɪʒn]n.视觉
visualize['vɪʒuəlaɪz]vt.形象，形象化
voice[vɔɪs]n.语音，声音
voltage['vəʊltɪdʒ]n.电压，伏特数
voluminous[və'lu:mɪnəs]adj.体积大的，庞大的
vulnerability[ˌvʌlnərə'bɪlətɪ]n.弱点，攻击
vulnerable['vʌlnərəbl]adj.易受攻击的

W

warehouse['weəhaʊs]n.仓库；vt.储入仓库
waterfall['wɔ:təfɔ:l]n.瀑布

wear[weə]vt.穿着，戴着
Web[web]n.万维网
website['websaɪt]n.网站
widespread['waɪdspred]adj.分布广泛的，普遍的
willingness['wɪlɪŋnəs]n.自愿，乐意
wire['waɪə]n.电线
wireless['waɪələs]adj.无线的
wordiness['wɜ:dɪnəs]n.累赘，冗长
wordy['wɜ:dɪ]adj.多言的，冗长的，废话连篇，唠叨的
workflow['wɜ:kfləʊ]n.工作流程
workstation['wɜ:ksteɪʃn]n.工作站
worm[wɜ:m]n.蠕虫
wrap[ræp]vt.遮蔽，隐藏
wraparound['ræpə,raʊnd]adj.包着的，包围的，环绕的

Z

Zurich['zjʊərik]n.苏黎世

附录B 词 组 表

a group of 一群，一组
a kind of... ……的一种
a range of 一系列，一些，一套
a sense of ... 一种……感觉
a variety of 多种的
a wide variety of 种种，多种多样
access point 访问点，接入点
active hub 有源集线器
active session 有效对话期间，工作时间
agile model 敏捷模型
analog signal 模拟信号
analog-to-digital conversion 模（拟）数（字）转换
antivirus software 防病毒软件
application program 应用程序
application service provider 应用服务提供商
application software 应用软件
applications software package 应用软件包
applied mathematics 应用数学
artificial general intelligence 通用人工智能
as the name suggests 顾名思义
assembly language 汇编语言
assembly line（工厂产品的）装配线，流水线
associate with … 同……联合
at the same time 同时；一起
away from 远离
back end 后端
base upon 根据，依据
batch program 批处理程序
be based on... 基于……
be categorized as ... 被分类为……
be characterized in ... 以……为特征
be concerned with 关心，关注，关于
be conflated with ... 与……混为一谈
be confused with … 与……混淆
be designed to ... 被设计为……
be divided into 被分为
be incorporated into ... 被并入……
be independent of 无关，不依赖，不取决于，不受……限制或制约
be instructed to 被指令做某事
be known as 被称为
be referred to as 被称作；被称为
be regarded as 被认为是
be restricted to 仅限于
be retrieved from... 从……收集，从……获取
be shielded from 与……隔离
be suited to 适合
be susceptible to … 对……敏感，可被……
be thought of 被认为
be transported up to ... 被输送到……
be viewed as 被看作是，视为
Bell Labs 贝尔实验室
big data 大数据
binary digit 二进制位，二进制数字
binary number 二进制数
bit string 位串

black box 黑箱
blackout blindfold 遮光眼罩
block storage 块存储
boolean value 布尔值
bus topology 总线拓扑
business decision 业务决策，商务决定
business opportunity 业务机会，商业机会
business process 业务流程
business process model 业务处理模型
by means of 通过，用……方法
cardboard goggles 纸板护目镜
carry out 执行，进行，完成
chat room 聊天室
Chemical Engineering 化学工程
child prodigy 神童
child table 子表
circuit-level gateway 电路层网关
classical music 古典音乐
client-side scripting language 客户端脚本语言
clock interrupt 时钟中断
cloud bridge 云桥
cloud computing 云计算
cloud desktop 云桌面
cloud operating system 云操作系统
cloud storage 云存储
cloud-based client operating system 基于云的客户端操作系统
code generation 代码生成
code portability 代码可移植性
code walk-through 代码走查
coding style 编程风格，编码风格
colocated cloud computing 同位云计算
command line interface 命令行接口；命令接口
competitive advantage 竞争优势
computer chip 计算机芯片
computer model 计算机模型
computer network 计算机网络
computer science 计算机科学
computer worm 计算机蠕虫
computer aided design 计算机辅助设计
computer aided process 计算机辅助过程
computerized parts inventory system 计算机零部件库房管理系统
computing platform 计算平台
conditional loop 条件循环
conditional statement 条件语句
confidence scoring schema 置信评分模式
consists of... 由……组成
consulting firm 咨询公司
consumption-based pricing model 基于消费的定价模型
content management system 内容管理系统
contract with … 与……签订合同
control unit 控制单元，控制器
convert … into … 把……转换为……
cooperative storage cloud 协同存储云
cost allocation 成本分摊，成本分配
credit card 信用卡
critical code 关键代码
customer churn 客户流失
customer relationship management as a service 客户关系管理即服务
data backup 数据备份
data classification 数据分类
data dictionary 数据字典
data field 数据域，数据字段
data file 数据文件
data flow 数据流
data link 数据链路
data mart 数据集市
data mining 数据挖掘
data model 数据模型
data oriented 数据导向
data protection 数据保护
data set 数据集
data stream 数据流
data structure 数据结构
data warehouse 数据仓库

database model 数据库模型
decision tree 决策树
deep learning algorithm 深度学习算法
delivery schedule 交付时间表
delve into 钻研，深入研究
development environment 开发环境
device driver 设备驱动程序
digital camera 数码相机
digital signal 数字信号
directory tree 目录树
discriminated union 可区分联合
disjoint union 分割联合
disk drive 硬盘驱动器
disk space 磁盘空间
distributed operating system 分布式操作系统
distributed storage 分布式存储
divide … into … 把……分为……
divide up 分割
doctoral degree 博士学位
dynamic router 动态路由器
dynamic systems development method 动态系统开发方法
early warning 预警
electrical circuit 电路
electromechanical device 机电装置，电气机械装置
electronic switch 电子开关
eliminating distraction 消除干扰
embedded operating system 嵌入式操作系统
embedded system 嵌入式系统
enterprise search 企业搜索
error checking 差错校验
event-driven system 事件驱动系统
eventually consistent 最终一致性
expert system 专家系统
export restriction 出口限制
external cloud 外部云
fail to… 未能……
fall into 分成，属于
fault tolerant 容错
faulty equipment 故障设备
fiber optic 光缆，光纤
fiber-optic cable 光纤电缆
figure out 弄清；计算出
file system 文件系统
financial institution 金融机构
fire door 防火门
flash memory 闪存
flat-file database 平面文件数据库
flight reservation system 机票预定系统
flight simulator 飞行模拟装置，飞行模拟器
floating point 浮点，浮点法
focus on 致力于，使聚焦于
for ... purpose 为了……目的
formal logic 形式逻辑
game theory 博弈论，对策论
genetic algorithm 遗传算法
glue together 胶合，黏合
government agency 政府机构
graphic card 图形卡
graphical interface 图形界面，图形接口
graphics rendering 图形渲染
graphics tablet 绘图板；图形输入板
hard disk 硬盘
hash table（散列）表
heath care 卫生保健
higher-level language 更高级的语言
high-level language 高级编程语言
high-performance computer 高性能计算机
host computer 主机
human intelligence 人类智能
hydrogen bomb 氢弹
hypertext link 超文本链接
identity management 身份管理
image recognition 图像识别
in response to... 对……做出反应
index file 索引文件
infrastructure as a Service 基础架构即服务
input device 输入设备

insist on 坚持
instant message 即时消息，即时报文
instead of 代替，而不是……
integrated circuit 集成电路
integrated development environments 集成开发环境
interact with ... 与……相互作用，与……交互；与……相互配合
interactive game 互动游戏
internal cloud 内部云
interpreted language 解释语言
investment decision 投资决策
iterative and incremental development 迭代和增量开发
iterative development 迭代开发
java virtual machine java 虚拟机
just-in-time compiler 即时编译器
keep an eye on 密切注视，照看
keep track of 跟踪
key-aggregate cryptosystem 密钥聚合密码系统
keylogger software 键盘记录软件
killer application 杀手级应用
knowledge discovery 知识发现
knowledge set 知识集
knowledge-driven decision 知识驱动决策
lawn mower 割草机，剪草机
light-weight iterative project 轻量级的迭代项目
load module 载入模块，装入模块
local drive 本地驱动器
local machine 本地计算机
logic gate 逻辑门
logical bus topology 逻辑总线拓扑
logical ring topology 逻辑环形拓扑
logical structure 逻辑结构
long integer 长整型
look up 查找
machine language 机器语言
machine vision 机器视觉
machine-based image processing 基于机器的图像处理
mainframe computer 大型计算机
map out 描绘出
market basket analysis 购物篮分析
mass storage 大容量存储器
media player 媒体播放器
medical image analysis 医学图像分析
memory cell 存储单元
memory space 存储空间
mesh topology 网状拓扑
microwave oven 微波炉
mobile device 移动设备，移动装置
modular programming 模块化编程
more or less 或多或少
motorcycle helmet 摩托车头盔
move along 向前移动
multiple task 多任务
multiprogramming operating system 多程序操作系统
multi-user operating system 多用户操作系统
natural disaster proof backup 防自然灾难备份
nearest neighbor 最邻近算法
negative number 负数
network component 网络元件
network layer 网络层
neural network 神经网络
not-so-distant future 不远的将来
numerical analysis 数字分析
object code 目标代码
object module 目标模块
object program 目标程序
object storage 对象存储
object-oriented programming 面向对象程序设计
on-demand service 按需服务
one at a time 一次一个
one-factor authentication 单因素认证，单身份验证
on-ramp system 入站匝道系统
open source 开源
operating system 操作系统
output device 输出设备
packet filter 包过滤
packet switching network 分组交换网
page description language 页面描述语言

parallel port 并行端口
parallel processing 并行处理
pass along 沿……走，路过
pass through 经过，通过
pattern recognition 模式识别
perimeter network 外围网络；非军事区网络
peripheral equipment 外围设备，外部设备
personal computer 个人计算机
personal experience 个人体验
physical star 物理星形
physical storage 物理存储，实际存储
piezoelectric sensor 压电传感器
plain old data structure 普通传统数据结构
plan-driven model 计划驱动模型，基于计划的模型
platform virtualization environment 平台虚拟化环境
plug into 把插头插入，接通
point-of-sale data 销售终端数据
pop up 弹出
position sensor 位置检测器，位置传感器
power supply 电源；供电
predictive analytic 预测分析
predictive model 预测模型
Princeton University 普林斯顿大学
print job 打印作业，打印任务
private cloud 私有云
program flow 程序流程
programming language 程序设计语言
project management 项目管理
protect from 保护
proxy server 代理服务器
public cloud 公有云
put up 提供，进行
quantum mechanics 量子力学
query language 查询语言
rapid prototyping methodology 快速原型法
raw data 原始数据，未加工的数据
reach out （使）伸出
real number 实数
real-time operating system 实时操作系统
receive side 接收方
recursive function 递归函数
reference model 参考模型
reinforcement learning 强化学习
relational database 关系数据库
relational database management system (RDBMS) 关系数据库管理系统
relationship marketing 关系营销
relative location 相对位置
requirements analysis 需求分析
requirements specification 需求规约书
response rate 响应率
retinal scan 虹膜扫描
ring topology 环形拓扑
risk management 风险管理
risk-driven spiral model 风险驱动的螺旋模型
road sign 交通标志，路标
root out 搜寻
routable address 可路由的地址
routable protocol 可路由协议
routine step 常规步骤
routing table 路由表
rule induction 规则归纳
rule set 规则集
scripting language 脚本语言
scroll wheel 滚轮，滚动轮，卷轴轮
secondary memory 辅助存储器
seemingly unrelated product 似乎无关的产品
self-driving car 自动驾驶汽车
semantic search 语义搜索
send side 发送方
sensitive data 敏感数据
sensory equipment 传感设备
sentient system 感觉系统
sentiment analysis 情感分析，倾向性分析
separates … into … 把……分为……
server-side scripting language 服务器端脚本语言
set in stone 一成不变
set theory 集合论

set up 建立；准备；安排
sift through 筛选
single-user operating system 单用户操作系统
smart phone 智能电话
social media 社交媒体
social network 社交网络
social setting 社会环境，社会场景，社会情境
software as a service 软件即服务
software design document 软件设计文档
software developer 软件开发人员
software development process 软件开发过程
software engineer 软件工程师
software engineering 软件工程
software testing 软件测试
sound card 音效卡，声卡
source code 源编码，源程序
source program 源程序
spam detection 垃圾邮件检测
special character 特殊字符
speech recognition 语音识别
spell out 讲清楚，清楚地说明
spiral model 螺旋模型
spread to 传到，波及，蔓延到
square root 平方根
star topology 星形拓扑
start with … 以……开始
state table 状态表
stateful firewall 状态防火墙，基于状态检测的防火墙
stateless firewall 无状态防火墙
static router 静态路由器
stereo sound 立体声
stock market 股市
storage area 存储区
storage space 存储空间
straight line 直线
striking building 引人注目的建筑物
strong AI 强人工智能
structured programming 结构化程序设计
supervised learning 监督学习
supervisor mode 管态
surround-sound loudspeaker 环绕声扬声器
system software 系统软件
systems development life cycle 系统开发生命周期
tagged union 标签联合
take a look into 看一看
take off 拿掉，取消
take place 发生，进行
talking point 话题；论题；论据
targeted marketing 目标市场
technological progress 技术进步
telephone cabling 电话线
telephone network 电话网
television program 电视节目
test data generation 测试数据生成
text file 文本文件
text translation 文本翻译
the duration of … 在……期间
the U.S. Constitution 美国宪法
thick coax 粗同轴电缆
thin Ethernet network 细线以太网，细缆以太网
third party 第三方
three-factor authentication 三因素认证；三重身份验证
time consuming 耗费时间的
time slice 时间片
time-based fee 基于时间收费
time-sharing operating system 分时操作系统
together with 和，加之
token ring network 令牌环网
top-down design model 自顶向下的设计模型
traffic signal 交通信号；红绿灯
transaction data 业务数据；交易数据；事务数据
translate … into … 把……翻译为……
transport layer 传输层
trial period 试用期
trigonometric function 三角函数
trip up 犯错误，绊倒，使摔倒
turn off 关闭
turn on 打开

turns … into … 把……转变为……
twisted pair 双绞线
twisted pair cable 双绞线
two-factor authentication 双因素认证；双重身份验证
unsupervised learning 无监督学习
user mode 用户态
virtual health assistant 虚拟健康助理
virtual machine image 虚拟机映像
virtual mode 虚（拟）模式
virtual personal assistant 虚拟个人助理
virtual reality 虚拟真实，虚拟现实
waterfall model 瀑布模型
weak AI 弱人工智能
Web developer 开发人员
Web Mining 网络挖掘
Web page 网页
welding visor 焊接面罩
WiFi hotspot WiFi 热点
wired Ethernet network 有线以太网
wireless communication 无线通信
word processing program 字处理程序
work together 合作，协作，一起工作
World Wide Web 万维网
wraparound headset 环绕式耳机

附录C 缩略语表

3D (3 Dimensions) 三维、三个维度、三个坐标
3GL (Third-Generation Language) 第三代语言
4GL (Fourth-Generation Language) 第四代语言
AI (Artificial Intelligence) 人工智能
AIaaS (Artificial Intelligence as a Service) 人工智能即服务
ALU (Arithmetic and Logic Unit) 算术逻辑部件
ANSI (American National Standards Institute) 美国国家标准协会
API (Application Program Interface) 应用程序接口
AR (Augment Reality) 增强现实
ASCII (American Standard Code for Information Interchange) 美国信息交换标准码
ASP (Active Server Pages) 活动服务器页面
ASP (Application Service Provider) 应用服务提供方
ATM (Asynchronous Transfer Mode) 异步传输模式
BASIC (Beginners All-purpose Symbolic Instruction Code) 初学者通用指令码
BDaaS (Big Data as a Service) 大数据即服务
BI (Business Intelligence) 商业智能
BIOS (Basic Input/Output System) 基本输入输出系统
BNC (Bayonet Nut Connector) 刺刀螺母连接器，同轴电缆接插件
BSD (Berkeley Software Distribution) 伯克利软件发布
B-tree (Binary tree) 二叉树
CaaS (Cloud as a Service) 云即服务
CAD (Computer-Aided Design) 计算机辅助设计
CAE (Computer-Aided Engineering) 计算机辅助工程
CASE (Computer-Aided Software Engineering) 计算机辅助软件工程
CAVE (Cave Automatic Virtual Environment) 洞穴式自动虚拟环境
CDN (Content Delivery Network) 内容分发网络
CD-ROM (Compact Disc，Read-Only-Memory) 只读光盘
COA (Cloud-Oriented Architecture) 面向云的体系结构
COAX (COAXial cable) 同轴电缆
COBOL (COmmon Business-Oriented Language) 面向商业的通用语言
COM (Component Object Model) 组件对象模型
COOP (Continuity Of OPeration) 运营连续性
CPU (Central Processing Unit) 中央处理器
CRM (Customer Relationship Management) 客户关系管理
DB (DataBase) 数据库
DBMS (DataBase Management System) 数据库管理系统
DISA (Defense Information Systems Agency) 国防信息系统局

DMZ (DeMilitarized Zone) 非军事区
DR (Disaster Recovery) 灾难恢复
DSS (Decision Support System) 决策支持系统
EAI (Enterprise Application Integration) 企业应用集成
EDA (Event-Driven Architecture) 事件驱动体系结构，面向事件的体系结构
EMI (ElectroMagnetic Interference) 电磁干扰
EPROM (Erasable Programmable Read-Only Memory) 可擦可编程只读存储器
FAQ (Frequently Asked Question) 常见问题
FORTRAN (FORmula TRANslation) 公式翻译语言
GDPR (General Data Protection Regulation) 通用数据保护条例
GNU GNU is Not Unix 的递归缩写
GSNW (Gateway Service for NetWare) NetWare 网关服务
GUI (Graphical User Interface) 图形用户界面，图形用户接口
HMD (Head-Mounted Display) 头戴式显示器
HOA (Hypermedia-Oriented Architecture) 面向超媒体的体系，超媒体驱动的体系结构
HTML (HyperText Markup Language) 超文本标识语言
I/O (Input/Output) 输入/输出
IA (Information Assurance) 信息保障
IaaS (Infrastructure as a Service) 基础设施即服务
ID (identification, identity) 身份，标识符
IEC (International Electrotechnical Commission) 国际电工委员会
IEEE (Institute for Electrical and Electronics Engineers) 电气和电子工程师学会
IP (Internet Protocol) 网际协议
IPC (interprocess communication) 内部进程通信
IPS (Intrusion Prevention System) 入侵防御系统
IPX (Internetwork Packet Exchange protocol) 互联网分组交换协议
ISDN (Integrated Service Digital Network) 综合业务数字网
ISO (International Organization for Standardization) 国际标准化组织
IT (Information Technology) 信息技术
JPEG (Joint Photographic Experts Group) 联合图像专家组
JSP (Java Server Pages) Java 服务器页面
Kbps (Kilobits per second) 千位每秒
LAN (Local Area Network) 局域网，本地网
LISP (LISt Processor) 列表处理语言
MAU (Multistation Access Unit) 多站访问接入单元
Mbps (Megabits per second) 兆位每秒
MiFi (Mobile wireless fidelity) 移动无线保真
MIMD (Multiple Instruction Multiple Data) 多指令多数据
MPU (Micro Processor Unit) 微处理器单元
NASA (National Aeronautics and Space Administration) 美国航空航天局
NAT (Network Address Translation) 网络地址转换
NDIS (Network Driver Interface Specification) 网络驱动器接口标准
NIC (Network Interface Card) 网络接口卡，网卡
NLP (Natural Language Processing) 自然语言处理
NT (network termination) 网络终端
OLTP (OnLine Transaction Processing) 联机事务处理
OO (Object-Oriented) 面向对象的
OOP (Object-Oriented Programming) 面向对象程序设计
OSI (Open System Interconnect) 开放式系统互联
OSS (Object Storage Service) 对象存储服务
PaaS (Platform as a Service) 平台即服务
PAD (Packet Assembler/Disassembler) 分组装拆器
PC (Personal Computer) 个人计算机
PDA (Personal Digital Assistant) 个人数字助理，掌上电脑
PII (Personally Identifiable Information) 个人身份信息
RAD (Rapid Application Development) 快速应用软件开发
RAM (Random Access Memory) 随机存取存储器
RDBMS (Relational DataBase Management System) 关系数据库管理系统
RJ (registered jack) 注册的插座
ROA (Resource-Oriented Architecture) 面向资源的体

系结构，资源驱动的体系结构
ROM (Read-Only Memory) 只读存储器
RPA (Robotic Process Automation) 机器人进程自动化
RPC (remote procedure call) 远程过程调用
SaaS (Software as a Service) 软件即服务
SAN (Storage Area Network) 存储域网
SDLC (Software Development Life-Cycle) 软件开发生命周期
SEPG (Software Engineering Process Group) 软件工程过程组
SIMD (Single Instruction Multiple Data) 单指令多数据
SMTP (Simple Mail Transfer Protocol) 简单邮件传输协议
SNA (Systems Network Architecture) 系统网络体系
SOA (Service-Oriented Architecture) 面向服务的体系结构，服务导向的体系结构
SPX (Sequences Packet Exchange) 顺序分组交换
SQL (Structured Query Language) 结构化查询语言
TB (TeraByte) 太字节
TCP (Transmission Control Protocol) 传输控制协议
TDI (Trandport Driver Interface) 传输驱动程序接口
TTL (Time To Live) 存活时间
UDP (User Datagram Protocol) 用户数据报协议
URI (Uniform Resource Identifier) 统一资源定位符
USASI (United States of America Standards Institute) 美国标准学会
USB (Universal Serial Bus) 通用串行总线
UTP (Unshielded Twisted Paired) 非屏蔽双绞线
VPN (Virtual Private Network) 虚拟专用网络
VR (Virtual Reality) 虚拟现实
VRML (Virtual Reality Markup Language) 虚拟现实标识语言
WAN (Wide Area Network) 广域网
WAP (Wireless Access Point) 无线接入点，无线访问点
WiFi (Wireless Fidelity) 基于 IEEE 802.11b 标准的无线局域网
XOR (Exclusive OR) 异或（逻辑运算）
XP (eXtreme Programming) 极限编程